Supply Chain Simulation

Francisco Campuzano · Josefa Mula

Supply Chain Simulation

A System Dynamics Approach for Improving Performance

Springer

Francisco Campuzano
Department of Business Management
Escuela Técnica Superior de Ingeniería Industrial
Technical University of Cartagena
Campus Muralla del Mar
30202 Cartagena (Murcia)
Spain
e-mail: francisco.campuzano@upct.es

Josefa Mula
Research Centre on Production Management and Engineering (CIGIP)
Department of Business Management
Universitat Politècnica de València
Campus de Alcoy
Plaza Ferrándiz y Carbonell, 2
03801 Alcoy
Spain
e-mail: fmula@cigip.upv.es

ISBN 978-0-85729-718-1 e-ISBN 978-0-85729-719-8
DOI 10.1007/978-0-85729-719-8
Springer London Dordrecht Heidelberg New York

British Library Cataloguing in Publication Data
A catalogue record for this book is available from the British Library

Translators: Helen Warburton and Antonio Maravilla Burgos (HyA Translations)

Cover design: eStudio Calamar S.L.

Printed on acid-free paper

Springer is part of Springer Science+Business Media (www.springer.com)

Contents

Chapter 1
Introduction to Supply Chain Simulation

1.1 Introduction

A supply chain is a network of facilities and distribution options that performs the following functions; the procurement of materials, transformation of these materials into intermediate and finished products; distribution of these finished products to customers. Supply chain management is a strategy through which the integration of these different functions can be achieved (Shapiro 2000).

Supply chain simulation implies operating a model that suitably represents a supply chain. Supply chain management can be carried out in the model should it be impossible, too expensive or impractical to do so in the real organization. Model performance may be studied and the properties relating to real supply chain performance may be deduced.

There are several reasons to simulate the supply chain. It could prove impossible or costly to observe certain processes in a real supply chain, for instance, sales in forthcoming years, etc. A supply chain can be too complex to describe it as mathematical equations. Even if a mathematical model was formulated, it could be too complex to obtain a solution by means of analytical techniques. It is feasible to study changes in a supply chain in a model and/or to verify analytical solutions. Supply chain simulation can provide a valuable idea about the most important variables and how they interact. It can also be used to experiment with new situations about which little or no information is available (uncertainty), and to check new policies and decision rules before risking experiments with the real supply chain.

This chapter discusses the use of analytical or simulation models to then describe the characteristics of simulation-based models. Next, it defines the main supply chain simulation objectives and highlights supply chain simulation method. Finally, this chapter addresses different supply chain simulation techniques.

F. Campuzano and J. Mula, *Supply Chain Simulation*,
DOI: 10.1007/978-0-85729-719-8_1, © Springer-Verlag London Limited 2011

1.2 Analytical or Simulation-Based Models?

Analytical models present a series of advantages that concisely describe the problem, provide a closed series of solutions, allow an easy assessment of the impact caused by changes in inputs on output measures, and offer the possibility of reaching an optimum solution. Their main drawbacks relate to the assumptions made to describe a system as they may not be very realistic and/or the mathematical formulae can be very complicated and interfere with finding a solution.

Simulation models can describe highly complex systems, and be used to either experiment with systems that still do not exist or experiment with existing systems without altering them (this may also be done using analytical methods provided the system is not highly complex). Among the drawbacks, one worthy of mention is that these models do not generate a closed set of solutions. Each change made in the input variables requires a separate solution or a series of runs. Complex simulation models may entail a long time to be constructed and run. Furthermore, model validation may prove a difficult task (that is, correspondence with the real system).

There are times when the combined use of both methods proves fruitful. The advantage of this mixed, or hybrid, approach is that analytical models are able to produce optimum solutions, whereas a suitable degree of realism and the accuracy of the system's description are reflected with simulation models. However, this combination has a disadvantage in that it requires a greater level of familiarity with analytical models, and also more skill than if using simulation models alone.

We refer readers to "A theory of supply chains" by Daganzo (2003) for an analytical approach of supply chain modeling. By summarizing lectures given at University College Berkeley, it highlights the connections among the traffic flow, queuing systems and supply chains.

This book describes the use of simulation-based models for supply chain modeling. The study on the supply chain will be done by means of simulation when one or several of the following conditions apply (Shannon 1975):

- The problem has no mathematical formulation.
- There is a mathematical model, but it has no analytical resolution methods.
- There is a model and methods, but the procedures are tedious, and simulation is simpler and less costly.
- When the aim is to observe a simulated history of the supply chain.
- When the aim is to experiment with a model before configuring the supply chain.
- It is impossible to experiment on the real supply chain.
- It is possible to experiment on the supply chain, but ethical reasons hinder this.
- When the aim is to observe very slow supply chain evolution by reducing the time scale.

1.3 Characteristics of a Simulation Model

Simulation is an important tool to explain how supply chain performance indicators react in the face of controllable factors and environmental factors. Thus, simulation is an experimental method. Experiments can be done with different input values and with several simulation model structures (representing various policies, etc.), considered a black box. Some methods do not treat the simulation model as a black box: perturbation analysis and score function (Spall 2003). However, these methods require more mathematical conditions being met and analysts being more mathematically sophisticated.

Simulation is widely used in practice as it does not require sophisticated mathematics. In supply chain simulation, a large number of responses is natural (service level, stocks, sales, etc.), and the balanced scorecard context (as in Kleijnen and Smits (2003)) discusses and frames these responses.

Simulation may offer an idea about the causes and effects of supply chain performance. What inputs significantly affect what outputs? Simulation can help understand causality, and it is a methodology that might not treat a supply chain as a black box. For instance, the Arena software (Kelton et al. 2004) could model individual events in great detail, such as the arrival of orders and machine failures.

A simulation model is characterized as being quantitative, mathematical and computer-based. It is also a dynamics model. It has at least one equation with at least one variable which refers to at least two different time points (for example, differential equations). It is not solved through a mathematical analysis. The temporal pathways of the dependent variables (outputs) are solved given not only the initial simulated system status, but also the values of the exogenous variables (inputs). Simulation does not offer a closed solution, but it allows us to observe what happens with output (sensitivity analysis) in terms of input values and model structures.

Below there is a list of the different guidelines indicated for simulation-based modeling:

- Not constructing a complicated model but a simple one that works.
- Understanding problem modeling to use a suitable technique.
- Models must be validated before applying them.
- A model must never be put under pressure to do, or be criticized for not doing, that which it has never been devised for.
- A model cannot be better than the information entering it.
- A model must never be considered literally.
- Models cannot replace decision makers.

1.4 Objectives of Supply Chain Simulation

There is a series of general supply chain simulation objectives.

- Generating supply chain knowledge. Using simulation to understand all or part of the supply chain, its processes and key problems.
- Developing and validating improvements. Simulation can be used to propose and simulate scenarios to improve the supply chain (what-if analysis).
- Reproducing and testing different decision-based alternatives. Determining a priori the level of optimization and robustness of a strategy without interrupting the real supply chain.
- Quantifying benefits. In general, simulation is important because it could help quantify the benefits resulting from the supply chain management supporting decision making at the strategic decision level: supply chain configuration (including (re)designing the supply chain), and at the tactical and operational decision levels: supply chain coordination (including the establishment of control policies values).

The specific supply chain simulation objectives will center, on the one hand, on the supply chain network design. This will imply logic design or the modeling and configuration of nodes (Hirsch et al. 1998), and the localization of nodes based on placing a supply chain node in a specific geographical area (Stefanovic et al. 2009). Melo et al. (2009) provide a literature review of facility location models in the supply chain management context. They concentrate their review on articles published in the last decade (including a few papers published in 2008) that go beyond location–allocation decisions (thus, they exclude simple single facility location and pure resource allocation models). Moreover, they only consider discrete models. They list several important issues that enable a facility location model to become compatible with supply chain network design requirements, and they survey the existing literature in relation to the following features: network structure and basis features, decision variables in supply chain network design, reverse logistics, other supply chain characteristics, supply chain optimization (performance measures and solution methodology) and applications.

On the other hand, simulation can act as a support for supply chain decision making (strategic, tactical and operational): rapid responses, collaborative planning, aggregated planning, forecasting demand, subcontracting third parties, etc. Some examples can be seen in Campuzano et al. (2010, 2011).

1.5 Types of Supply Chain Simulation

From the methodology perspective, Kleijnen (2005) distinguishes four types of simulation problems for supply chains: validation and verification; sensitivity or the what-if analysis, which provides a list, or screening, of the most important factors in simulation models of supply chains; optimizing the critical control factors; and robustness, risk, or the uncertainty analysis.

A robust solution obtains the important factors controlled by management by considering the noise created by non controllable environmental factors. Supply chain management distinguishes between robustness and flexibility: a flexible supply chain can react to a changing environment by adapting its operations. A robust supply chain ensures it has a set design, and can still adapt to many changes in its environment. To deal with these four methodological problems, a variety of simulation types and techniques may be used.

Kleijnen and Smits (2003) distinguish among the following simulation types for supply chains.

- Simulation using a spreadsheet.
- Systems dynamics. It can demonstrate the bullwhip effect (see Chap. 3). It is also useful for chain management, BPR (Business Process Reengineering).
- Simulation of systems dynamics with discrete events. It can quantify service levels, particularly under uncertainty by focusing on an analytical simulation.
- Business games. They can educate and train users since players are active participants in the simulated world. Besides, they can be involved in investigation to study the effects of the qualitative factors (i.e., the decision system) on benefits, etc. They are also suitable for a distributed virtual environment.

The type of simulation that must be used will depend on the problem to be solved by the problem in each specific case. Next, the main characteristics of each simulation type are highlighted.

1.5.1 Spreadsheet-Based Simulation

The introduction of spreadsheets has made production processes modeling popular in companies (Plane 1997; Powell 1997). This kind of simulation is quite credible for directors and managers. However, assessing the results of proposing these simulation models with spreadsheets may prove too simple and unreal.

1.5.2 System Dynamics

Forrester (1958, 1961) proposes a methodology for the simulation of dynamic models: industrial dynamics; which is the origin of system dynamics (Sterman 2000). Industrial dynamics is a quantitative approach that studies the characteristics of the information feedback of industrial systems to understand how the organizational structure, amplification (in politics), and time delays (in decisions and actions) interact to influence the company's success. Companies are considered from a methodology perspective as systems with six types of flows:

information, orders, materials, money, staff and equipment. The theoretical supply chain studied by Forrester (1961) is formed by four levels: retailer, wholesaler, distributor and manufacturing. How these nodes react before the deviations between the current inventory levels and the objective inventory levels is examined. One conclusion is that common-sense strategies could amplify end consumers' demand fluctuations along the supply chain. Later, Lee et al. (1997a, b) identify how the sales-related demand distortion due to the Forrester effect further amplified, which was owing to the following effects that may even show simultaneously in the supply chain: order sizing, product price fluctuation, rationing and lack of finished products. The combination of these four elements leads to the amplification of variance in product demand. This amplification of demand, which increases the further we are from the end customer and the more we enter the supply chain, is known as the bullwhip effect, and can be used to measure supply chain management efficiency.

In general, the main objective of system dynamics is to understand the structural causes that bring about the behavior of a system (Sterman 2000). Some examples of input flows in the supply chain are production and sales. Some examples of output flows in the supply chain are stocks, fill rate and work in process (WIP). Systems dynamics assumes that control is carried out by varying the ratio of the variables (for instance, production and sales) which changes flows (and therefore stocks). It is also based on the feedback principle, i.e., a manager compares an objective value for a metrics with the real value and takes corrective actions, if required.

This is a rigorous method for a qualitative description of exploring supply chains as far as their processes, information, strategies and organizational limits are concerned. It facilitates modeling and the qualitative simulation analysis to design and control the supply chain structure. It also facilitates experiments with supply chains. It does not require detailed information or exact data on relationships. It focuses on the dynamic performance of the combination of feedback loops.

Some simulation works relating to the supply chain are based on systems dynamics, and can be found in Ashayeri and Keij (1998); Angerhofer and Angelis (2000); Beamon (1998); Otto and Kotzab (2003); Tesfamariam and Lindberg (2005); Pierreval et al. (2007) and Campuzano et al. (2010).

1.5.3 Discrete Events Systems Dynamic

Discrete event systems dynamics are an important type of supply chain simulation with two main characteristics: represents individual events (the arrival of an order from a customer, etc.) and incorporates uncertainty (customers' orders arrive at random time points, machines break down at random repair times, etc.). The work of Law and Kelton (2000) on discrete event simulation based on the Arena ® Software is worthy of mention, as iss the review of works about supply chain simulation by Banks et al. (2002).

1.5.4 Business Games

It is difficult to model human performance. One solution involves allowing directors to operate with the supply chain (SC), and its simulated environment by management or a business game. The objectives here may be of an educational and research kind. Through their 'Lean Leap Logistic Game', Holweg and Bicheno (2002) demonstrate how a participative simulation model is used to not only reveal the dynamics of a supply chain from the automobile sector, but to also model possible improvements.

Kleijnen and Smits (2003) distinguish two subtypes of business games: strategic games which include several teams of players that represent the companies competing in the simulated world. One well-known example is the Beer Game which illustrates the bullwhip effect (Sterman 1989, Sterman 2000). Then there are operational games which include a single team (with one or several players) who interact with the simulation model during a few rounds in real time. Some examples are the games used for production planning training. Kleijnen and Smits (2003) provide more references on SC-related games.

1.6 Techniques for Supply Chain Simulation

Local simulation is the most widely used supply chains simulation technique found in the literature. It consists in a single simulation model being run on a single computer (Terzi and Cavalieri 2004). This can be done by resorting to a specific supply chain simulation software: i2 (Padmos et al. 1999), IBM Supply Chain Simulator (Bagchi et al. 1998), SDI Industry Pro (Siprelle et al. 1999), SCGuru (www.promodel.com), LOCOMOTIVE (Hirsch et al. 1998), Supply Solver (Schunk 2000), among others. Then, there is general purpose software: Arena (www.arena.com), ModSim (www.modsim.org), Promodel (www.promodel.com), Vensim (www.vensim.com), Dynamo, ITHINK, Powersim and Stela, among others. Although general purpose software offers more flexibility, it is more complex.

The parallel simulation technique consists in running simulation programs on multi-processor calculation platforms. However, the distributed simulation technique consists in running simulations on computers that are geographically interconnected by a network. Both cases involve running a main simulation model made up of several sub-simulation models, which are run in a distributed fashion on a large number of calculation stations.

The distributed and parallel simulation technique consists in several simulation models that are run on several processors (computers and/or multi-processors) in a parallel and distributed manner (Terzi and Cavalieri 2004). This technique is based on the collaboration concept in which each model co-participates in a single simulation run and as a single decision maker in a federate environment. There are four main reasons to resort to this technique:

- It is possible to divide a large simulation into more models and to run it in less time.
- To preserve the geographic distribution, running a distributed and parallel simulation on distributed computers will enable the creation of virtual worlds with a good number of participants who are physically located in different plants.
- Under the distributed and parallel simulation paradigm, it is possible to include different existing models and different simulation tools in a single environment without having to adopt a platform, a common language and having to rewrite models.
- Should a processor fail, other processors can continue the simulation.

It is feasible to design a supply chain simulation model and to carry it out traditionally as a single model by reproducing all the nodes or by using more integrated models (one per node) to be run in a parallel fashion in single cooperation simulation. Evidently, this form of cooperation simulation requires more complex information and communication technologies than local simulation does.

1.7 Conclusions

This Introduction chapter justifies the need for simulation models for supply chains, and reviews the main forms of supply chain simulation and existing techniques for this very purpose.

At this point, systems dynamics is selected in general, and the Vensim© DSS simulation software in particular, as the basis for this book to study supply chain dynamics problems. In relation to systems dynamic, we agree with Sterman (2000) and Amir (2005) about its effectiveness to model dynamic business systems, in this case supply chains. As regards the selected software, the reason it employs a modeling approach is that it combines systems dynamics concepts and simulation with discrete events to represent supply chain events and uncertainty in detail, and to subsequently analyze their performance based on their structure and the existing causal relationships among their components. Local simulation is used to model the proposed example problems.

References

M.S. Amir, *Industrial viewpoint. Can systems dynamics be effective in modelling dynamic business systems? Bus.* Process. Manag. J. **11**, 612–615 (2005)

B.J. Angerhofer, M.C. Angelis, *System dynamics modelling in supply chain management: research view*, ed. by Joines et al. In: Proceedings of the 2000 Winter Simulation Conference, Orlando, pp. 342–351 (2000)

J. Ashayeri, R. Keij, Global business process re-engineering: A system dynamics based approach. Int. J. Oper. Prod. Manage. **18**, 817–831 (1998)

S. Bagchi, M. Buckley, M. Ettl, G. Lin, *Experience using the IBM Supply Chain Simulator. Proceedings of the 1998 Winter Simulation Conference*, Washington, pp. 1387–1394 (1998)

J. Banks, S. Buckley, S. Jain, P. Lendermann, *Opportunities for simulation in supply chain management*. ed. by Yücesan et al. In: Proceedings of the 2002 Winter Simulation Conference, San Diego, pp. 1652–1658 (2002)

B.M. Beamon, Supply chain design and analysis: Models and methods. Int. J. Prod. Econ. **55**, 281–294 (1998)

F. Campuzano, A. Lisec, A. Guillamón, Assessing the impact of prices fluctuation on demand distortion within a multiechelon supply chain. Promet **23**, 131–140 (2011)

F. Campuzano, J. Mula, D. Peidro, Fuzzy estimations and system dynamics for improving supply chains. Fuzzy. Set. Syst. **156**, 1530–1542 (2010)

C.F. Daganzo, *A theory of supply chains* (Springer, Heidelberg, 2003)

J.W. Forrester, Industrial dynamics. A major breakthrough for decision makers. Harvard. Bus. Rev. **36**, 37–66 (1958)

J.W. Forrester, *Industrial Dynamics* (MIT Press/Wiley, New York, 1961)

B.E. Hirsch, T. Kuhlmann, T.J. Schumacher, Logistics simulation of recycling networks. Comput. Ind. **24**, 31–38 (1998)

M. Holweg, J. Bicheno, Supply chain simulation–A tool for education, enhancement and endeavor. Int. J. Prod. Econ. **78**, 163–175 (2002)

W.D. Kelton, R.P. Sadowski, D.T. Sturrock, *Simulation With Arena*, 3rd edn. (McGraw-Hill, Boston, 2004)

J.P.C. Kleijnen, Supply chain simulation tools and techniques: A survey. Int. J. Simulat. Pro. Model **1**, 82–89 (2005)

J.P.C. Kleijnen, M.T. Smits, Performance metrics in supply chain management. J. Oper. Res. Soc. **54**, 507–514 (2003)

A.M. Law, W.D. Kelton, *Simulation modelling and analysis*, 3rd edn. (McGraw-Hill, New York, 2000)

H.L. Lee, P. Padmanabhan, S. Whang, The bullwhip effect in supply chains. Sloan Manage. Rev. **38**, 93–102 (1997a)

H.L. Lee, P. Padmanabhan, S. Whang, Information distortion in a supply chain: The Bullwhip Effect. Manage. Sci. **43**, 543–558 (1997b)

M.T. Melo, S. Nickei, F. Saldanha-da-Gama, Facility location and supply chain management-A review. Eur. J. Oper. Res. **196**, 401–412 (2009)

A. Otto, H. Kotzab, Does supply chain management really pay? Six perspectives to measure the performance of managing a supply chain. Eur. J. Oper. Res. **144**, 306–320 (2003)

J. Padmos, T. Hubbard, S. Duczmal, S. Saidi, *How i2 Integrates Simulation in Supply Chain Optimization*. Proceedings of the 1999 Winter Simulation Conference, Phoenix, pp. 1350-1355 (1999)

H. Pierreval, R. Bruniaux, C. Caux, A continuous simulation approach for supply chains in the automotive industry. Simulat. Model Pract. Theory **15**, 185–198 (2007)

D.R. Plane, How to build spreadsheet models for production and operations management. OR/MS Today **24**, 50–54 (1997)

S.G. Powell, Leading the spreadsheet revolution. OR/MS Today **24**, 8–10 (1997)

D. Schunk, *Using simulation to analyze supply chain*. Proceedings of the 2000 Winter Simulation Conference, Orlando, pp. 1095–1100 (2000)

R.E. Shannon, *Systems Simulation: The Art and Science* (PrenticeHall, Englewood, 1975)

J.F. Shapiro, *Modeling the Supply Chain* (Duxbury Press, Pacific Grove, 2000)

A.J. Siprelle, D. Parsons, R.A. Phelps, *SDI Industry Pro: Simulation for Enterprise Wide Problem Solving*. Proceedings of the 1999 Winter Simulation Conference, Phoenix, pp. 241–248 (1999)

J.C. Spall, *Introduction to Stochastic Search and Optimization: Estimation, Simulation and Control* (Wiley, Hoboken, New Jersey, 2003)

J.D. Sterman, *Business Dynamics: Systems Thinking and Modeling for a Complex World* (McGraw-Hill Higher Education, New York, 2000)
J.D. Sterman, Modeling managerial behavior: misperceptions of feedback in a dynamic decision making experiment. Manage. Sci. **35**, 321–339 (1989)
D. Stefanovic, N. Stefanovic, B. Radenkovic, Supply network modelling and simulation methodology. Simulat. Model Pract. Theory **17**, 743–766 (2009)
S. Terzi, S. Cavalieri, Simulation in the supply chain context: A survey. Comput. Ind. **53**, 3–16 (2004)
D. Tesfamariam, B. Lindberg, Aggregate analysis of manufacturing systems using system dynamics and ANP. Comput. Ind. Eng. **49**, 98–117 (2005)

Chapter 2
Conceptual Framework for Supply Chain Simulation

2.1 Introduction

Supply chain management involves the integration of key business processes from the end user through original suppliers which provide products, services and information with added value for customers and other interested parties (Lambert and Cooper 2000).

This chapter presents a conceptual framework that describes the nature of the most relevant interrelations and elements in designing supply chain management, and one that can act as a basis for the subsequent creation of different supply chains models. In accordance with Lambert and Cooper (2000), the supply chain (SC) conceptual framework consists in three closely interrelated elements: the SC structure (network of companies), SC business processes and SC components.

This chapter reviews each interrelated element making up the SC conceptual framework as proposed by Lambert and Cooper (2000), and which this book takes as its conceptual basis. Next, the main procedures to consider for SC simulation are covered. Finally, a learning activity is proposed to complement the contents of this chapter.

2.2 The Supply Chain Network Structure

The SC network structure comprises all the companies participating in a production chain, services ranging from raw materials to end consumers, and the connections among them (thanks to which, commercial activities or business processes are carried out). According to Lambert et al. (2000), this structure is made up of the central company (or the control company) and several of its links (suppliers and customers). The dimensions to consider include SC length and the number of suppliers and customers at each level. It is interesting to note that the SC does not appear as such, but better resembles tree branches where roots and

F. Campuzano and J. Mula, *Supply Chain Simulation*,
DOI: 10.1007/978-0-85729-719-8_2, © Springer-Verlag London Limited 2011

branches symbolize a network (Cooper et al. 1997), which is the reason why it seems odd to find only one company participating in a chain. With all this in mind, a dilemma is faced: how many of these branches and roots need managing? The commonest factors determining the amount of companies which must be managed in the SC concept are, according to Lambert et al. (2000), product complexity, number of suppliers and raw material availability.

Managers and administrators suggest that not all the links throughout the SC must be strictly coordinated and integrated into management because the level of relationship among links differs vastly. In SC management, it is necessary to select the most appropriate level of society for each particular link (Lambert et al. 2000). Evidently, the most appropriate relationship is that which is the most important for the company (Cooper et al. 1993).

In order to learn and know how the SC network is outlined, Lambert et al. (2000) suggest analyzing three structural aspects of the network:

- Supply chain members.
- The network's structural dimensions.
- The different types of links making up the processes.

2.2.1 Identifying Supply Chain Members

In order to determine the network structure, it is necessary to identify the SC members. They must be classified by level, and how vital they are for company success must be assessed. Integrating and coordinating all the links of the process could, in the vast majority of cases, be counterproductive, complex and even impossible (Cooper et al. 1997).

Supply chain members include all the companies or organizations with which the central company acts reciprocally, directly or indirectly through its suppliers or customers from the point of origin to the point of sales. Nevertheless to make a complex network more manageable, it is important to distinguish primary members from support members (Davenport 1993).

Primary SC members are those autonomous companies or strategic commercial units that carry out activities with added value, operational or management activities in commercial processes which generate specific output for a particular customer or market. Conversely, support members are companies that simply supply resources, knowledge and tools for an SC's primary members. For instance, support companies include transport companies, banks that lend money, the owner of a building offering storage space, and the companies that supply production equipment, prepare printed commercial leaflets, etc.

One company can do both activities: primary and support. Likewise, one company can undertake primary activities relating to a process and to support activities in relation to another process. For example, when a manufacturer buys production equipment from a supplier that has been designed according to a joint

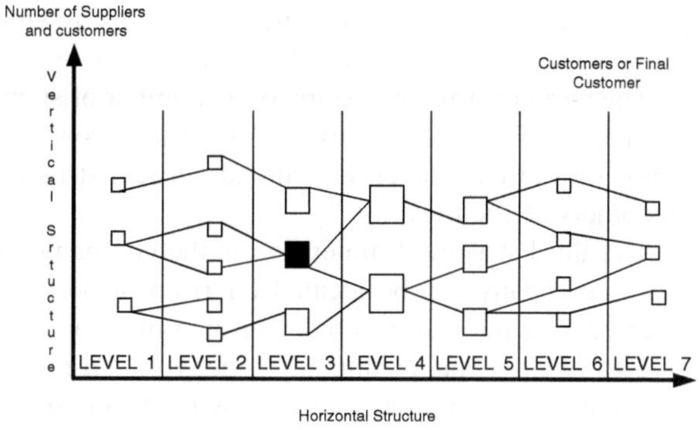

Fig. 2.1 Network structural dimensions. Source: Lambert et al. (2000)

development project of a given product, this equipment's operation is ensured for this new article. In this way, the supplier becomes a primary member of the manufacturer's development process. However, it also becomes a support member because supplying the equipment itself does not add value to process performance, even though the equipment does add value.

It is worth pointing out that the difference between SC primary and support members is not so obvious in all cases. Nonetheless, the definition previously provided at least offers a reasonable administrative simplification which covers the essential aspects as to who can be considered an important SC member.

The definition of primary and support members enables the definition of the SC's point of origin and point of sale. In general terms, there is no primary member in the SC's point of origin as they are all considered support members.

2.2.2 Network Structural Dimensions

The three essential network structural dimensions for the description, analysis and administration of an SC are (Lambert et al. 2000):

- The horizontal structure.
- The vertical structure.
- The horizontal position within the chain.

Lambert et al. (2000) establish the former and subsequent levels to the chain's main company (Fig. 2.1 represents it with a blackened square) to analyze the chain and its relationships, thus establishing the network structural dimensions.

In Fig. 5.1, the black-filled square represents the chain's main company (a primary member), while the other empty squares depict its support companies.

The horizontal structure refers to the number of levels in the SC. This can be large or small according to the number of existing levels. For instance, the network structure for the automobile industry is excessively large. Vehicle parts are made in several places by a large amount of suppliers around the world which send their products to the main automobile subsystems assembly centers, and which subsequently travel long distances for final vehicle assembly.

The vertical structure refers to the number of suppliers or customers represented at each level. One company can have a narrow vertical structure with very few companies at each level, or a broad vertical structure with many suppliers and/or customers at each level.

The third structural dimension is the company's horizontal position within the SC. A company can be located far from or near the initial supply source, or far from or near the end customer, or at some place between these SC extremes.

The chain's main company's network size can make its relationships with customers and suppliers difficult, with which spin-off activities tend to be promoted to fulfill the level of appropriate service for all the chain's customers (Stern et al. 1995).

Finally, it is worth distinguishing among the various types of links in the SC:

- Management links: important for the integration and management performed by the central company.
- Monitored links: not as critical for the central company as the former.
- Non management links: not sufficiently critical.
- Non member links: they do not belong to the SC, but their decisions can affect its performance.

2.3 Supply Chain Business Processes

Business processes are those activities that generate a specific output of value to the customer. Systems dynamics-based supply chain simulation implies formerly identifying the key business processes in the SC.

Supply chain management requires information to flow constantly for the purpose of producing the most appropriate flow of assets toward customers. It is important to remember that due to the SC management approach having the customer as its basis, accurate and appropriate information of processes is required so that the companies comprising the SC respond to the frequent changes and fluctuations in demand. Having controlled the changes in the customer's demand pattern, industrial processes and the supplier's action are basic in SC efficiency.

The relevant business processes identified by the Global Supply Chain Forum (GSCF) members within the SC are as follows (Lambert et al. 2000):

1. Management of relationships with the customer

The first step in SC management integration is to identify the customers or group of customers that may be considered essential or important in the company's business mission. It will be vitally important to specify the level of service to be accomplished with these customers by identifying and eliminating the causes generating demand variability.

Administration of the relationships with customers involves making performance assessments, which analyze the level of service provided to customers and the profitability of these customers.

2. Customer service management

Thanks to a greater interrelation with the production area and the organization's distribution systems, the customer service department can provide information about its commitments with deliveries, dates and product availability, etc., in real time. In an SC system, this department's functions include orientating the customer as to the use of the commercialized products.

3. Demand management

Hewlett-Packard's experience in the SC indicates that two inventory types must be distinguished: the essential and the variable (Davis 1993). An essential inventory includes the products being manufactured and the goods being moved from one place to another through the commercialization channels. In a variable inventory, the stocks resulting from the fluctuations in manufacturing processes, supply and demand are identified. Customer demand is the main source of variability and is made up of irregular patterns. Since customers' orders are unforeseen, demand management is a key element in efficient SC management.

During the demand management process, the customer's requirements must be balanced with the company's supply capacity in an attempt to determine the exact amount and when to buy by using demand forecasting techniques. To lower the level of uncertainty, demand management systems use points of sales and the most important customers' databases which improve the efficiency of the physical flow of merchandises throughout the SC.

4. Fulfilling orders

The key to efficient SC management lies in meeting customers' requirements. For this reason, an efficient process that integrates manufacturing, distribution and transport plans is needed to fulfill orders. To achieve this, pacts must be agreed on with the key SC members, especially with transport companies with a view to fulfilling the customer's requirements, while reducing total distribution costs at the same time. The aim must be to develop a management process from the supplier to several customers segments.

5. Manufacturing flow management

In traditional companies, manufacturing flows management follows a common process: producing, storing and delivering end products to the distribution system in accordance with historical demand forecasts. With this manufacturing outline, products are manufactured in accordance with a strict production schedule. Yet one common characteristic of this type of systems is that they present unnecessary and excessive inventories which normally generate high costs.

In the processes-oriented management approach currently employed for SC management, the product is manufactured according to the customer's requirements.

Manufacturing processes can be made flexible to respond to market changes by installing dynamic systems which can be adapted to different product characteristics such as mass customization. It consists in the design, production, marketing and delivery of customized products and services on a mass basis (Pine 1992). Some mass customization strategies are: supply chain management, modular product design, virtual enterprise, the web, best-of-breed information and communication technologies, and agile manufacturing. Thus, orders are processed with JIT (just-in-time) systems in minimum amounts and with the manufacturing priorities defined by the delivery date (Schonberger 1980). This approach, which emphasizes simple control systems, has brought about changes to the manufacturing process with shorter cycle times and improved customer service.

6. Supplies or purchases

The main function of the supplies or purchases process is to develop strategic plans with suppliers that support the manufacturing flow management process and new products development. Similarly, suppliers are classified in this stage according to the various dimensions, such as their contribution and importance for the organization. As part of this process, long-term strategic alliances are developed with a small group of suppliers for the purpose of obtaining mutual profit that is protected within the "win–win" type relational models.

The philosophy of this process intends to involve important suppliers in the earliest stages of the design cycle to significantly cut manufacturing cycle and product delivery times (de Treville et al. 2004).

Nowadays, the purchases area consolidates and improves its operation by information and communication technologies to transfer the information relating to its requirements. This rapid communication among chain elements cuts time and costs in terms of the transactions that the process results in.

7. Product development and commercialization

In supply chain management, suppliers and their customers join to develop new products for the purpose of cutting commercialization times. When products' life cycles shorten, these are launched in the market for shorter times to remain competitive. Based on this outline, commercialization development and processes managers are obliged to (Lambert et al. 2000):

- Coordinate with the customer service area to identify the end customer's requirements.
- Select materials and suppliers along with the supplies department.
- Develop technologies to facilitate production and the integration of flows into the SC, thus accomplishing the best product/market combination.

8. Returns

Suitable management of the returns channel, otherwise known as reverse logistics, as a business process offers the chance to achieve a considerable competitive advantage from the sales perspective (Clendein 1997); Lambert et al. (2000)

point out that efficient returns management helps identify opportunities to improve productivity and to discover new projects. This perhaps applies to some cases, but reverse logistics is a partial solution whose ultimate purpose is to eliminate inefficiencies and unnecessary controversies which emerge during SC activities. Indeed it is hoped that, with time, SC elements should do away with all kinds of outlines from a previously set up quality decision-making platform that allows correct communication and operation to avoid returns. However, it is important to have an explicit logistics scheme to provide improved returns management. Besides, returns will be defined by the product type in each case.

2.4 Supply Chain Components

Supply chain management components are variables by which business processes are integrated and managed by means of a SC.

Lambert et al. (2000) propose the following components for a successful SC:

- Planning and control.
- Work structure, which defines how tasks and activities are to be done.
- Organizational structure of an individual company or the SC.
- Structure of installations for products flow. A network of installations to supply, produce and distribute throughout the SC.
- Structure of installations for information flow, which defines the type of information and how often is it updated.
- Management methods based on corporate philosophy and management techniques.
- Structure of leadership and authority. Power, or lack of it, in the SC can affect other SC members' level of commitment.
- Shared awards and risks.
- Attitude and culture.

2.5 Supply Chain Simulation Procedures

This section defines supply chain simulation procedures.

The first procedure to be carried out refers to understanding industry's characteristics, as well as the SC's business and planning processes. Some examples of these planning processes to be simulated are:

- Planning sales and demand. Generating the customer process and defining demand forecasting planning.
- Supply chain planning or supply chain network design. Assigning production and distribution resources with capacity and supply constraints, e.g., Supply Solver (Schunk and Plott 2000).

- Inventory planning. Designing and managing storage policies and procedures for raw materials, the current inventory and end products; e.g., SCGuru from Promodel (www.promodel.com).
- Production planning and scheduling. Designing and managing the production process (scheduling and acquiring raw materials, production process design and scheduling, designing and controlling materials flow); e.g., SDI Industry Pro (Siprelle et al. 1999).
- Planning transport and distribution. Simulating distribution centers, location of warehouses, and planning transport in terms of resources, dates and costs, e.g., IBM Supply Chain Simulator (Bagchi et al. 1998).

These processes interact to produce an integrated SC. The design and management of these processes determine the SC's extension or scope as a unit to fulfill the required operation objectives.

The next stage is to design a simulation scenario. More often than not, it is not reasonable to model all the details, so it is necessary to center on the problem areas. This is when the decision variables in SC simulation are defined. Here are some examples according to Beamon (1998):

- Scheduling production/distribution. Scheduling amounts (how much) and dates (when) to be produced and/or distributed.
- Inventory levels. Determine the amount and location of storing all raw materials, intermediate and end products.
- Number of stages (levels). Determine the number of SC stages. This involves increasing or decreasing the vertical integration levels of the chain by combining (or eliminating) or separating (or adding) stages.
- Assigning a distribution-customer center. Deciding which distribution centers will serve which customer.
- Assigning the plant-product. Determining which plants will manufacture what products.
- Buyer–supplier relationships. Determining and developing critical aspects of the buyer–supplier relationship.
- Specifying the product differentiation step. Deciding on the product manufacturing process step in which the product should be differentiated (or specialized).
- Number of product types maintained in the inventory. Determining the number of different products types that will make up the end products inventory.

It is also necessary to collect all the data required for SC simulation. Below is a feasible example of the data required for SC simulation (Table. 2.1):

One important procedure in supply chain simulation is establishing suitable outcome indicators and their objectives.

An outcome indicator, or series of outcome indicators, is used to determine the efficiency and/of effectiveness of a system, or to compare it with an alternative system. It is also used to design systems by determining the decision variables values that make the outcome indicators more advantageous. Beamon (1998) distinguish between qualitative and quantitative outcome indicators.

Table. 2.1 An example of data requirements for supply chain simulation

Area	Data required
Manufacturing process and times	Manufacturing process data (processing time, lead times, preparation times, number of machines in each process, alternative routes).
	Schedule data (shifts, vacations, preventive maintenance).
	Machinery data (number of machines, mean failure times, mean repairs time, alternative resources, preventative maintenance time).
	Lists of materials.
Inventory control policies	Safety stock levels, order points.
	Inventory level of end products, raw materials and intermediate products.
	Location of the stock in the production plant.
Supplies and logistics	Supplier's supply time. Supplier's lot size. Supplier's capacity. Supplies horizon. Supplies time.
Demand	Service date. Priority. Start and end dates. Demand pattern.
Policies/strategies	Control policies for orders. Service policies.

Qualitative outcome indicators are used when there are no direct numerical measures, and when some aspects can be quantified. Here are a few examples:

- Customer satisfaction (internal or external) (Christopher, 1998):
  - Pre-transaction satisfaction. Associated with the service elements that are given a priori product purchase.
  - Transaction satisfaction. Associated with the service elements implied in the physical distribution of products.
  - Post-transaction satisfaction. Associated with the support provided while using the product.
- Flexibility. The extent to which the SC can respond to random fluctuations in the demand pattern.
- Integration of materials and information flow. Magnitude with which all the SC functions communicate information and transport materials.
- Effective risk management. Extent of minimizing inherent SC risks.
- Suppliers' performance. This determines the consistency with which suppliers deliver raw materials to production plants in terms of delivery dates and good conditions.

Quantitative outcome indicators can be numerically described, as the examples below indicate:

- Objectives based on cost/profit:
  - Minimizing costs. This is the most widely used. The total cost of the SC, of the business units or of specific stages is minimized.
  - Maximizing sales. This maximizes the amount of sales in either monetary units or sold units.
  - Maximizing the gross margin. It maximizes income less costs.

  - Minimizing the investment made in inventory. It minimizes the costs of inventories policies (including product and maintenance costs).
  - Maximizing return on investment. It maximizes the ratio between net profit and invested capital.

- Objectives based on the customer's response:
  - Maximizing the fill rate. It maximizes the fraction or orders of customers served on time.
  - Minimizing delayed demand. It minimizes the length of time between a promised delivery date and the real delivery date.
  - Minimizing the lead time. It minimizes the length of time required since a product has started to be produced until it has been completely processed.
  - Minimizing duplicated operations. It minimizes the number of business operations carried out by more than one business organization.

Readers are referred to Otto and Kotzab (2003) for an extension of supply chain performance metrics.

Subsequently, it is necessary to define the conditions of finalizing simulation in order to avoid the transitional state and to stabilize the model. Some such conditions refer to:

- Checking that negative system loops predominate over positive system loops. Should it be a case of simulating demand management within the SC, inventory levels should not spiral without control. This means that resupplying the order would be erroneously calculated and the system would never leave the transitory state.
- Checking that the variables representing on-hand inventory do not present periods with negative values during the simulation period (this is only possible for the net inventory, i.e., the available inventory less pending orders).
- Should variables that measure system performance be introduced, it is necessary to ensure that the simulation horizon is long enough for these variables to show reliable results. Fundamental performance trends can only be seen in a sufficiently lengthy time scale. Sometimes, it should not be forgotten that the results of certain policies are not optimum because the decision-making time horizon was too short or because a system perspective was lacking when the problem was being considered.

Finally, it will be necessary to contemplate SC policies/strategies to be assessed with the developed simulation model. The various policies when it comes to launching resupply or production orders, supply times or production times (depending on the level being modeled), production capacity (number of operators, number of hours per product), the forecasting techniques employed, the management policies for orders (FIFO, LIFO, etc.), all enable different simulation scenarios to be considered, and the choice of that which best fits the proposed business objectives (storage costs, levels of service, etc.).

2.6 Learning Activity

In order to complement this chapter, readers are recommended to study the articles by Archibald et al. (1999), Bagchi et al. (1998), Holweg and Bicheno (2002) and Otto and Kotzab (2003).

2.7 Conclusions

This chapter reviews the conceptual framework for the supply chain management proposed by Lambert et al. (2000), which this book takes as the conceptual basis for supply chain models. Thus, three structural aspects of the supply chain network are reviewed: SC members, structural network dimensions and the different kinds of links composing the supply chain processes to be examined. Based on this, a series of procedures required to simulate a supply chain is proposed. Finally, a learning activity has been added to complement the contents of this chapter.

References

G. Archibald, N. Karabakal, P. Karlsson, Supply Chain vs. *Supply Chain: Using Simulation to Compete Beyond the Four Walls.* Proceedings of the 1999 Winter Simulation Conference, Phoenix, vol 2, pp. 1207–1214 (1999)

S. Bagchi, M. Buckley, M. Ettl, G. Lin, *Experience Using the IBM Supply Chain Simulator.* Proceedings of the 1998 Winter Simulation Conference, Washington, pp. 1387–1394 (1998)

B.M. Beamon, Supply chain design and analysis: Models and methods. Int. J. Prod. Econ. **55**, 281–294 (1998)

M. Christopher, *Logistics and Supply Chain Management: Strategies for Reducing Costs and Improving Service* (Prentice Hall, Financial Times, London, 1998)

J.A. Clendein, Closing the supply chain loop: reengineering the returns channel process. Int. J. Logist. Manage. **8**, 75–83 (1997)

M.C. Cooper, J.T. Gardner, Good business relationships: More than just partnerships or strategic alliances. Int. J. Phys. Distrib. **23**, 14–20 (1993)

M.C. Cooper, L.M. Ellram, J.T. Gardner, A.M. Hanks, Meshing multiple alliances. J. Bus. Logist. **18**, 67–89 (1997)

T.H. Davenport, *Process Innovation: Reengineering Work Through Information Technology* (Harvard Business School, Boston, 1993)

T. Davis, Effective supply chain management. Sloan Manage. Rev. **34**, 35–46 (1993)

S. de Treville, R.D. Shapiro, A.P. Hameri, From supply chain to demand chain: The role of lead time reduction in improving demand chain performance. J. Oper. Manag. **21**, 613–627 (2004)

M. Holweg, J. Bicheno, Supply chain simulation—a tool for education, enhancement and endevaour. Int. J. Prod. Econ. **78**, 163–175

D.M. Lambert, M.C. Cooper, Issues in supply chain management. Ind. Market Manag. **29**, 65–83 (2000)

A. Otto, H. Kotzab, Does supply chain management really pay? Six perspectives to measure the performance of managing a supply chain. Eur. J. Oper. Res. **144**, 306–320 (2003)

B.J. Pine II, *Mass Customization. The New Frontier in Business Competition* (Harvard Business School Press, Cambridge, MA, 1992)
R.J. Schonberger, *Japanese Manufacturing Techniques. Nine Hidden Lessons in Simplicity* (Free Press, New York, 1980)
D. Schunk, B. Plott, *Using Simulation to Analyze Supply Chain.* Proceedings of the 2000 Winter Simulation Conference, Orlando, vol 2, pp. 1095–1100 (2000)
A.J. Siprelle, D. Parsons, R.A. Phelps, *SDI Industry Pro: Simulation for Enterprise Wide Problem Solving.* Proceedings of the 1999 Winter Simulation Conference, Phoenix, pp. 241–248 (1999)
L.W. Stern, A. El-Ansary, *Marketing Channels*, 5th edn. (Prentice Hall, Englewood Cliffs, 1995)

Chapter 3
Bullwhip Effect in Supply Chains

3.1 Introduction

The bullwhip effect refers to the scenario in which orders to the supplier tend to present larger fluctuations than sales to the buyer, and the resulting distortion increasingly amplifies upstream in a supply chain.

Jay Forrester and Jack Burbidge are pioneers in our modern knowledge of the supply chain.

Forrester (1958) analyzes a traditional supply chain and observes how a small change in a customer's demand pattern amplifies as it flows through distribution, production and replenishment processes. At each supply chain level, this deviation is amplified upstream to the supply chain as replenishment orders. This effect is known as the Forrester effect and is one of the indicators of inefficient supply chain management. Forrester (1961), known as the "father" of supply chain design, established the bases of research into supply chain performance and characteristics in his book entitled "Industrial Dynamics" which presents the so-called Forrester effect, the precursor of the bullwhip effect. According to Forrester, this amplification is owing to the problems arising due to there being no zero lead times, and also due to the inaccurate forecasts made by the different supply chain members when faced with demand variability.

Burbidge (1961, 1983) presents the so-called PBC (Period Batch Control), which develops the 'five golden rules to avoid bankruptcy':

- Rule 1. Only manufacture those products that can be quickly dispatched and invoiced to customers.
- Rule 2. Only manufacture those components in one period that are needed in the next period.
- Rule 3. Minimize the materials processing time.
- Rule 4. Use the shortest planning period that can be efficiently managed.
- Rule 5. Only accept deliveries from suppliers in small batches and when required for processing or assembly.

F. Campuzano and J. Mula, *Supply Chain Simulation*,
DOI: 10.1007/978-0-85729-719-8_3, © Springer-Verlag London Limited 2011

Subsequently, Towill (1997) integrates Forrester and Burbidge's concepts to develop a series of improved communication and materials flow practices in the supply chain, known as FORRIDGE, which is based on the following 4 + 1 principles:

- Control system principle. It is the need to select the most suitable control system to accomplish the user's objectives.
- Time-compressing principle. All supply chain activity should be carried out in the minimum time required to accomplish task objectives. So it is necessary to eliminate added time without value, or the system's idle time, and to deliver what is required on time.
- Information transparency principle.
- Levels elimination principle. There should be the minimum number of appropriate levels for the accomplishment of the supply chain's goals, and there should be minimum stocks in the right place and at the right time.
- Synchronization principle. All the events are synchronized in such way that orders and deliveries are visible at all the discrete time points.

Wikner et al. (1992) add a sixth principle to the FORRIDGE model, the multiplier principle. Orders are multiplied between products manufacturers and their equipment suppliers. If a products manufacturer renews its machines for a 10 year cycle, it could choose to increase its capacity by 10% a year, and manages to double its machines orders; that is, a multiplier of 10–1.

Lee et al. (1997a) identify how the sales-related demand distortion, due to the Forrester effect, amplifies even more because of the following effects, which may even show simultaneously in the supply chain: order lot sizing, rationing, lack of finished products and product price fluctuation. The combination of these four elements leads to the amplification of product demand variance. This amplification of demand, which increases the further we are from the end customer and the more we enter the supply chain, is called the bullwhip effect. Lee et al. (1997b) study demand information flow and propose a theoretical framework for evaluating the effects of systematic information distortion through the supply chain. The authors assume that: (1) past demands are not used for forecasting; (2) re-supply is infinite with a fixed lead time; (3) there is no fixed order cost; and (4) the purchase cost of the product is stationary over time. This ideal situation is useful as a starting point to analyze the bullwhip effect in a supply chain.

This chapter reviews the bullwhip effect concept in the supply chain using different examples in the literature. Then, the main causes of the bullwhip effect are examined, as are the strategies to neutralize them proposed by Lee et al. (1997b). This chapter also presents the commonest forms of measuring the bullwhip effect. Finally, it relates how the structure of different supply chain types influence the bullwhip effect.

3.2 Examples of the Bullwhip Effect

Holmström (1997) analyzes the bullwhip effect in a supply chain of the grocery industry, Towill and McCullen (1999) study the bullwhip effect in a clothes supply chain, while Daganzo (2003) unveils the core causes of the bullwhip effect and describes control methods for eliminating all instabilities without increasing supplier costs.

However, Sterman (1989) provides the best illustration of the bullwhip effect, who investigates how human errors affect the dynamics of a system through its well-known business game, the so-called *Beer Game*, from the performance science perspective.

The beer distribution game is a role-playing simulation of an industrial production and distribution system performed at the MIT (Massachusetts Institute of Technology) to introduce management students to the systems dynamics and simulation concepts. The supply chain comprises four sectors: retailer, wholesaler, distributor and manufacturer. A person manages each sector without being able to communicate with the others. Each sector sends its demand (represented by cards) to its supplier on a weekly basis. There are backlogged deliveries and orders are received in each stage. These represent the time required to receive, process, send and deliver orders. The aim of this game is to minimize total costs by appropriately managing inventories to face demand uncertainty. Players must maintain their inventory as low as possible to avoid backlogged demand. The inventory is ordered and the delivery time is variable. Typical sessions involve 3–8 teams with four players. Between 36 and 50 weeks are simulated in 90 min sessions. Information circulates throughout the chain, although most players have access only to the information located immediately up-or downstream which the company provides them. Without a clear vision of the end customer's demand, many companies can rely only on the information they have access to, which is often distorted by other participants in the chain. Thus, information becomes distorted as it is transmitted among the various links forming the chain. In the example provided by Sterman (1989), large fluctuations dominate orders and the inventory, and a mean of 21 weeks is needed to cover the initial inventory levels. All in all, the bar's inventory levels wane, followed in sequence by a drop in the wholesaler's, distributor's and manufacturer's inventories. As inventories fall, orders tend to increase. An effective inventory (inventory less backlogged demand) generally becomes significantly negative, indicating that sectors have backlogs. The maximum average backlog stock is 35 boxes, which takes place between weeks 20 and 25. Inventory levels emerge as additional products are produced and delivered. Quite often, the inventory considerably exceeds the initial inventory levels. The average inventory peak is 40 boxes, which takes place between weeks 25 and 30. Orders drop as quickly as the inventory increases. The cause of the bullwhip effect is, therefore, both the lack of transparent information across the supply chain companies and the uncontrolled increase or decrease in the size of orders, which have nothing to do with real demand. Figure 3.1 reflects the bullwhip effect in the beer game.

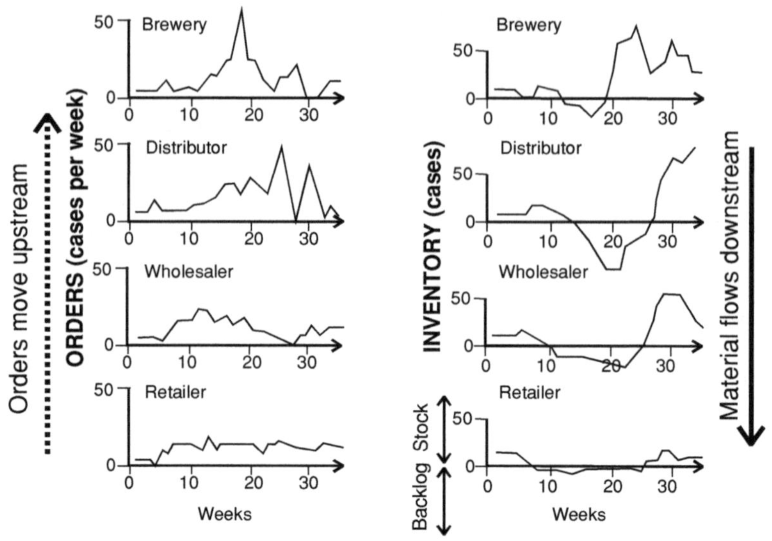

Fig. 3.1 The bullwhip effect in a supply chain. Source: Sterman (1989)

3.3 The Four Causes of the Bullwhip Effect

This section summarizes the four main causes of the bullwhip effect as analyzed by Lee et al. (1997a): processing demand and lead times that are not null, or the Forrester effect; order lot sizing or the Burbidge effect; rationing and scarcity of finished goods, or the Houlihan effect; prices variations or the promotion effect.

Processing the demand signal and non null lead times has, in the past, been called "Demand Amplification" or the "Forrester effect" after Forrester (1961) investigated the problem in many real supply chains and demonstrated it by simulation (DYNAMO). Sterman's limited rationing (1989) also dealt with the Forrester effect, a terminology that is common to the field of psychology to describe players whose behavior in decision-making seems rational, but is sub-optimum. Processing the demand signal is understood as the practice of adjusting demand forecasts and, as a result of this practice, parameters are adjusted to the inventory replacement rule. Processing the demand signal is the main contributing factor of the bullwhip effect. As safety stock contributes to the bullwhip effect, fluctuation becomes more significant when the lead times among suppliers lengthen.

Lot sizing is also known as the Burbidge effect (Burbidge 1991). It refers to the practice of managing orders downstream the supply chain (or in the manufacturing process) into lots to gain economies of scale in lot changing activities (machine configurations, or managing and receiving an order). This is often the result of either an EOQ calculation (Economic Order Quantity) or a similar technique. Periodic replenishment amplifies demand variability and contributes to the bullwhip effect. Optimizing transport is one of the causes of lot sizing. Towill (1997) discusses the problems that arise in detail.

Rationing and scarcity of finished goods (rationing and gaming), or the Houlihan effect, was examined by Houlihan (1987), who acknowledges that when scarcity, or unserved orders, occur in traditional supply chains (encouraged by suppliers to stimulate demand), customers increase their orders. This implies more demand in the production system, which inevitably leads to more unsatisfactory deliveries. Then customers increase their safety stocks, which distorts the demand signal, giving way to the bullwhip effect.

Prices variations, or the promotion effect, refer to the practice of offering products at lower prices to stimulate demand. The presumption of elastic demand temporarily increases demand ratios when customers make good use of this chance to buy in anticipation or increase their stocks. Yet this has a serious impact on supply chain dynamics because demand drops when the price no longer includes a discount. This gives the perception of having to offer more discounts to stimulate demand again.

Understanding the causes behind the bullwhip effect can help find strategies to mitigate it. Lee et al. (1997b) propose the following coordination mechanisms: shared information, supply chain alignment and operational efficiency; these relate with the four main causes of the bullwhip effect. Among the strategies to adopt, it is worth highlighting the following:

- Avoiding many demand forecast updates. To achieve this, different supply chain structures are proposed:
  - Demand data are available at all the supply chain levels (EDI, Internet, etc.), and this would be an EPOS-type (Electronic Point of Sales) supply chain structure.
  - Demand and inventories data are available at all the supply chain levels, and this would be a VMI-type (Vendor Managed Inventory) supply chain structure.
  - Direct sales will be made to the end consumer and this will be an e-shopping-type supply chain structure.
- No lot sizing:
  - JIT replenishment is an effective way to mitigate the effect by using smaller lot sizes or a more frequent replenishment system. The reason for large lot sizes may be due to costs relating to the order and to transport. Therefore, it is possible to resort to the use of external logistics operators.
- Stable prices:
  - Reduce the frequency and level of discounts in prices.
  - EDLP (Everyday Low Price) strategies.
  - CRP (Continuous Replenishment Program) strategies.
  - ABC (Activity-based Costing) systems.
- Eliminate rationing products situations:
  - Share information about capacities and inventories.
  - Work with customers to advance the production of seasonal sales.

- Other strategies:
  - Include information systems.
  - Define new organizational relationships.
  - Implement new measures and incentives systems.

Readers may refer to Swaminathan and Tayur (2003) for further information about the above.

Finally, it is worth stressing that the bullwhip effect generates additional costs in terms of insufficient or excessive capacity in the plant, employees' contract/dismissal, poor customer service levels, excess stock, obsolescence, poor suppliers' delivery and poor public image, among others.

3.4 Measuring the Bullwhip Effect

Geary et al. (2003) distinguish five different approaches to measure the bullwhip effect: operational research, the filter theory, the control theory, systems dynamics and ad-hocacy.

Operational research formulates the problem as a differential equation. The mathematical solution attempts to minimize a cost function subject to operation conditions. In many cases, the ratio between orders variance and demand variance is employed (Chen et al. 2000):

$$Bullwhip = \frac{\sigma_O^2/\mu_O}{\sigma_D^2/\mu_D} = \frac{\sigma_O^2}{\sigma_D^2} \tag{3.1}$$

This formula is based on the variation coefficient, that is, the dispersion statistics used in statistics to compare distributions from variability. It describes the existing ratio between variance and the arithmetic mean. A ratio higher than one indicates variance amplification, and below one implies an isolation effect. The mean could be eliminated from the formula provided distributions are equally distributed.

Fransoo and Wouters (2000) measure the bullwhip effect in a food supply chain in four different ways: individual product for specific sales, individual product for all sales, aggregated product for individual sales, and aggregated product for aggregated sales. They measure the bullwhip effect at a particular level in a multi-level supply chain as the quotient of the coefficient of the demand variance generated by this level, and as the coefficient of the demand variance received by this level:

$$Bullwhip = \frac{C_{out}}{C_{in}} \tag{3.2}$$

where

$$C_{out} = \frac{\sigma(O_n(t, t+T))}{\mu(O_n(t, t+T))} \tag{3.3}$$

and

$$C_{in} = \frac{\sigma(D_n(t, t+T))}{\mu(D_n(t, t+T))} \tag{3.4}$$

The filter theory expresses the problem in the frequency domain where value criteria are established based on the amplitudes of the "message", the "noise" or the "disturbances" spectrum (Towill and Del Vecchio 1994).

The control theory models the problem in a transformed manner and centers on the system's structure to initially ensure stability and to then form the desired response. To go about this, it uses the following: flow diagrams, block diagrams, z-transforms, control laws, and frequency and simulation response graphs.

Tsypkin (1964) demonstrated that the squared sum of the response to systems' impulse equals the noise bandwidth and the output variance divided by input variance:

$$Bullwhip = \frac{\sigma_O^2}{\sigma_D^2} = \frac{1}{\pi}\int_0^{\pi} |F^*(j\bar{w})|^2 d\bar{w} = \frac{W_N}{\pi} = \sum_{n=0}^{\infty} f^2(n) \tag{3.5}$$

The squared impulse method always produces numbered results rapidly. In some cases, the squared impulse method produces closed analytical forms. Sometimes the Parsevel relation can be used to calculate the noise bandwidth after using Tustin's transform. The noise bandwidth always produces numbered results and provides additional views on system performance.

Another way of measuring the bullwhip effect is the proposal of Dejonckheere et al. (2002) based on the following replenishment rule that "smooths" orders by amending certain parameters:

$$O_t = \hat{D}_t(T_a) + \frac{1}{T_n}(TNS_t - NS_t) + \frac{1}{T_w}(DWIP_t - WIP_t) \tag{3.6}$$

This manufacturing order was proposed by Towill (1982) and was later updated by John et al. (1994), Disney et al. (1997), and by Disney and Towill (2001).

Where $\hat{D}_t(T_a)$ is the demand forecast using simple exponential smoothing; T_a is the smoothing parameter used in the forecast; TNS_t is the target net stock level; NS_t is the net stock in period t; that is, the total inventory of the manufactured products less pending orders or orders still to be delivered; $DWIP_t$ is the desired work in process (WIP) level; WIP_t is the WIP in period t; T_n *and* T_w are smoothing parameters. T_n represents the inventory gain from the order command. T_w represents the gain in WIP from the order command. $TNS_t - NS_t$ represents the error in the net inventory in t. $DWIP_t - WIP_t$ represents the error in the work underway at time t.

Using this rule, slight adjustments can be made to the amount ordered as a response to changes in demand. It has the potential to smooth patterns of orders. The effect of the order smoothing proposed for the different T_n *and* T_w values is to diminish the bullwhip effect to a greater or lesser extent; in other words, amplifying orders according to the order's degree of smoothing.

In exponential smoothing, smoothing constant α can be represented as:

$$\alpha = \frac{1}{1 + T_a} \tag{3.7}$$

$$DWIP_t = T_p\bar{D}$$

To simplify, if $T_i = T_n = T_w$ and T_p is the lead time, then

$$DWIP_t = T_p\bar{D} \tag{3.8}$$

and (3.6) can be represented by

$$O_t = \bar{D} + \frac{1}{T_i}(TNS_t - NS_t) + \frac{1}{T_i}(DWIP - WIP_t) \tag{3.9}$$

Using Tsypkin's relation (1964) and a z-transform model, Disney and Towill (2002) derive the relationship among the bullwhip effect, smoothing parameters and production time:

$$Bullwhip = \frac{\sigma_O^2}{\sigma_D^2} = \frac{2T_a^2 + 3T_i + 2T_p + 2(T_i + T_p)^2 + T_a(1 + 6T_i + 4T_p)}{(1 + 2T_a)(T_a + T_i)(-1 + 2T_i)} \tag{3.10}$$

If $T_a = \infty$, then the expected long-term demand value can be used as a forecast, and (3.10) will be formulated as (3.11). This may be used if independent demand patterns are taken and are identically distributed; that is, a stationary demand pattern. If these assumptions are not valid, then T_a will differ from infinity, and forecasted demand will change from period to period.

$$Bullwhip = \frac{1}{2T_i - 1} \tag{3.11}$$

In (3.11), the bullwhip effect is independent of production time. If $T_i < 1$, then the variance of the order command amplifies in relation to demand, which implies the presence of the bullwhip effect. If $T_i = 1$, then a demand search strategy is adopted, meaning that the bullwhip effect does not exist. If $T_i > 1$, it represents order smoothing.

Likewise, Dejonckheere et al. (2002) obtain the net stock amplification (*NSAMP*), which also has an effect on the customer service level:

$$NSAmp = \frac{\sigma_{NS}^2}{\sigma_D^2} = 1 + T_p + \frac{2T_a^2(-1 + T_i)^2 + T_i(1 + T_p)^2 + T_a(1 + T_p)(1 + (-1 + 2T_i)T_p)}{(1 + 2T_a)(T_a + T_i)(-1 + 2T_i)} \tag{3.12}$$

For $T_a = \infty$, then:

$$NSAmp = 1 + T_p + \frac{(T_i - 1)^2}{2T_i - 1} \quad (3.13)$$

If $T_i = 1$, then $NSAmp = 1 + T_p$. Si $T_i > 1$ or $T_i < 1$, and *NSAmp* will increase. *NSAmp* is always greater than $1 + T_p$. As *NSAmp* contains a production time component and a smoothing component, the reduction of T_p will lower the *NSAmp* value.

Systems dynamics, the approach taken as the basis of this book, makes simulations through cause-and-effect diagrams and flow diagrams (Forrester 1961; Sterman 2000). This approach has been habitually employed to measure and reduce the bullwhip effect in supply chains. Wangphanich et al. (2010) develop a simulation approach based on system dynamics modeling and an adaptive network-based fuzzy inference system for quantifying and reducing the bullwhip effect in a multi-product, multi-stage supply chain. The model comprises three groups of variables which influence the bullwhip effect, namely the structure of a supply chain network, supply chain contributions (ordering process in a regular situation or when a supplier has a promotion or shortage gaming) and supply chain performances (the number of defects and the ordering lead time). The supply chain of a beverage company was selected to validate and demonstrate the proposed model's flexibility. Campuzano et al. (2010) evaluate the performance of fuzzy estimations of demand instead of exponential smoothing for demand forecasts in a two-level, single-item, multi-period supply chain. A system dynamics model with fuzzy demand estimations was constructed for supply chain simulation. Fuzzy numbers were used to model fuzzy demand estimations. A numerical example was used to show how the bullwhip effect and the amplification of inventory variance can be effectively reduced.

The ad-hocacy approach is based on practical experience to observe the amplification caused by a level, and to carry out actions to redesign the chain with a view to eliminating the level and assessing the improvement generated based on bullwhip knowledge (Holmström 1997).

3.5 Supply Chain Structure and the Bullwhip Effect

Disney et al. (2004) propose analyzing the bullwhip effect by studying five logistic supply chain structures. These structures are based on the use of information and communication technologies: traditional supply chain, reduced supply chain, e-shopping supply chain, and the EPOS (Electronic Point of Sales) and VMI supply chains.

In the traditional supply chain, each member (or level) sends an order to its supplier which delivers what is required. In this way, each member receives information about demand from the immediate downstream level and transmits it as orders to the immediate upstream level; in other words, there is access only to information about what the customer requires to place its own order. Therefore, the decision to place an order is made in two different time periods. Orders are made from a demand forecast, a component to calculate inventory levels and a component to calculate the level of scheduled receipts. This is the way followed in the beer game (Sterman 1989) and is reflected in industrial practice.

The main characteristic of a reduced supply chain is that one member or more of the traditional supply chain is/are eliminated. This is the case of chains that do not use retailers, such as direct sales of articles to end consumers on the Internet (Springer.com). The manufacturer can base its production orders on the end consumer's sales and can deliver directly to the end consumer. In this way, there are no intermediaries and the supply chain has a single level. In other cases, another firm can be involved in selling a range of products from several manufacturers. Thus, the supply chain may have two levels.

In an e-shopping supply chain, the factory receives the products or services demand directly from the end customer (which is the case of Dell Computers). This is also known as a traditional single-level supply chain.

The main characteristic of supply chains which employ the EPOS system is that the whole supply chain has access to data about current sales to the end consumer. In this way, each member knows the real products demand that the end customer requests in each period. The entire supply chain uses the end consumer's sales data for demand forecasts. Nonetheless, all the members have to deliver an amount ordered by their customer. At any rate, the different forecasting methods, and making use of opportunities to purchase raw materials at low prices, can lead to strange orders being placed that distort information and lead to the bullwhip effect. However, the bullwhip effect diminishes when all the EPOS supply chain members base their demand forecasts on the final consumer's demand more than they do on their sales. It is worth stressing that, according to Dejonckheere et al. (2003), the bullwhip effect can be reduced in correctly designed supply chains.

The VMI-type supply chain is characterized by there being an agreement between the distributor and the retailer (or between the manufacturer and the distributor). The distributor knows not only the end consumer's sales (receives information about them), but also the retailer's inventory and it manages this level. Moreover, the inventory is always within a level agreed upon by the two parties. VMI is a supply chain configuration where the end consumer's sales and the customer's inventory level are known by the supplier to set production and distribution objectives. Both the customer and supplier in the VMI agree on a work protocol in which the supplier is free to exploit the customer's stock level. There is only one decision point to order. To provide stock availability, the objective inventory level in the system changes temporarily, which adds variation to the orders ratio. The VMI evolves toward CPFR (Collaborative Planning, Forecasting and Replenishment) (Holmstrom et al. 2002), which includes demand planning. Danese et al. (2004) establish that CPFR-type mechanisms favor the generation of inter-company coordination mechanisms for the purpose of supporting the planning processes in the supply chain. Some benefits of collaborative systems are foreseeable, such as lower operational costs and more efficient processes (Li et al. 2007).

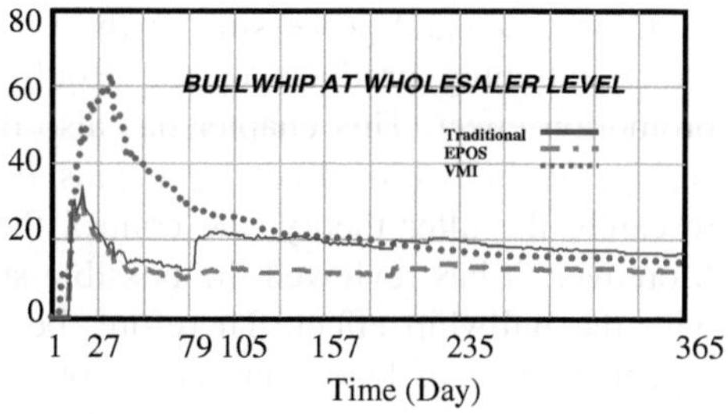

Fig. 3.2 Reducing the bullwhip effect at the wholesaler level. Source: Campuzano and Mcdonnell (2008)

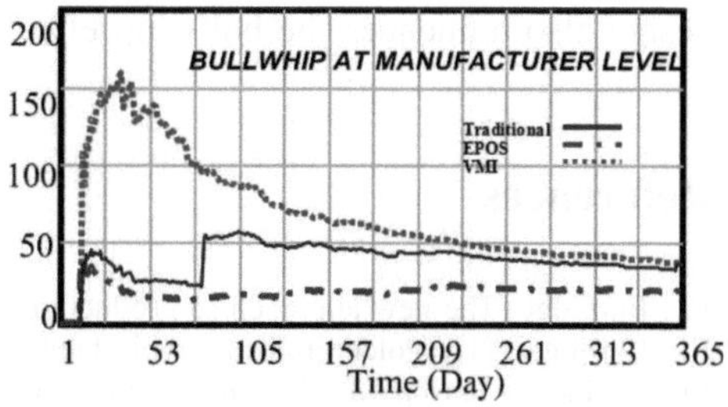

Fig. 3.3 Reducing the bullwhip effect at the manufacturer level. Source: Campuzano and Mcdonnell (2008)

Both VMI and CFPR have evolved from a traditional supply chain and avoid retailers, wholesalers and other distribution centers from disappearing, and manage to diminish the bullwhip effect. Disney and Towill (2001) relate the four main causes of the bullwhip effect (demand processing or the Forrester effect, lot sizing or the Burbidge effect, rationing products or the Houlihan effect, and prices variations or the promotion effect) with the bullwhip effect in a traditional and a VMI-type supply chain. They conclude that VMI completely avoids two bullwhip effect problems (the Houlihan and Burbidge effects). The Forrester effect is similar in both cases, but only 50% of a traditional supply chain's inventory levels is required for a VMI-type supply chain. Finally, VMI is able to better face prices promotions with an approximate 30% less increase in the bullwhip effect.

Figures 3.2 and 3.3 now compare the bullwhip effect for different types of supply chain structures (traditional, VMI and EPOS), specifically at the wholesaler and manufacturer levels. These figures originate from the study conducted by Campuzano and Mcdonnell (2008) on the use of collaborative structures to reduce the bullwhip effect. This study simulates several multi-level supply chains.

3.6 Conclusions

This chapter has reviewed the bullwhip effect concept that distorts demand information which spreads through the entire supply chain. With this in mind, the main causes of bullwhip effect, as identified by Lee et al. (1997b), have been examined: demand processing or the Forrester effect, lot sizing or the Burbidge effect, rationing products or the Houlihan effect and prices variations or the promotion effect. This chapter has also described the commonest forms of measuring the bullwhip effect by means of five different approaches: operational research, the filter theory, the control theory, systems dynamics and ad-hocacy. Moreover, it has reviewed the possible supply chain structures and their relation with the bullwhip effect. Therefore, because the traditional supply chain has a higher number of levels and does not exchange demand and/or inventory information, it will notice a greater bullwhip effect than those chains with a lower number of levels or which exchange information, such as the reduced type, e-shopping, EPOS or VMI supply chains. Nonetheless, adequate supply chain design also influences the bullwhip effect to a greater or lesser extent.

References

J.L. Burbidge, The new approach to production. Prod. Engineer. **40**, 769–784 (1961)

J.L. Burbidge, Five golden rules to avoid bankruptcy. Prod. Engineer. **62**, 13–14 (1983)

J.L. Burbidge, *Period Batch Control (PBC) with GT. The Way Forward from MRP* (BPCIS Annual Conference, Birmingham, 1991)

F. Campuzano Bolarín, L. Mcdonnell Ros, Reducing the impact of demand process variability within a multi-echelon supply chain. ICFAI. J. Supply. Chain. Manag. **5**, 7–21 (2008)

F. Campuzano, J. Mula, D. Peidro, Fuzzy estimations and system dynamics for improving supply chains. Fuzzy. Set. Syst. **156**, 1530–1542 (2010)

F. Chen, Z. Drezner, J.K. Ryan, D. Simchi-Levi, Quantifying the bullwhip effect in a simple supply chain: The impact of forecasting, lead-times and information. Manage. Sci. **46**, 436–443 (2000)

C.F. Daganzo, *A Theory of Supply Chains* (Springer, Heidelberg, 2003)

P. Danese, P. Romano, A. Vinelli, Managing business processes across supply networks: the role of coordination mechanisms. J. Purch. Supply. Manag. **10**, 165–17 (2004)

J. Dejonckheere, S.M. Disney, I. Farasyn, F. Janssen, M.R. Lambrecht, D.R. Towill, W. Van de Velde, (2002) in *Production and Inventory Control: The Variability Trade-Off*. Proceedings of the 9th EUROMA, Copenhagen

J. Dejonckheere, S.M. Disney, M.R. Lambrecht, D.R. Towill, Measuring and avoiding the bullwhip effect: A control theoretic approach. Eur. J. Oper. Res. **147**, 567–590 (2003)

S.M. Disney, M.M. Naim, D.R. Towill, Dynamic simulation modelling for lean logistics. Int. J. Phys. Distrib. **27**, 174–196 (1997)

S.M. Disney, M.M. Naim, A. Potter, Assesing the impact of e-business on supply chain dynamics. Int. J. Prod. Econ. **89**, 109–118 (2004)

S.M. Disney, D.R. Towill, (2001) in *The Impact of Vendor Managed Inventory on the Bullwhip Effect in Supply Chains*. Proceedings of the 16th International Conference on Production Research, Prague, Czech Republic

S.M. Disney, D.R. Towill, (2002) in*A Robust and Stable Analytical Solution to the Production and Inventory Control Problem via A Z-Transform Approach*. Proceedings of the 12th International Working Conference on Production Economics, Igls, Austria, Vol. 1, pp.37-47

J.W. Forrester, Industrial dynamics. A major breakthrough for decision makers. Harvard Bus Rev **36**, 37–66 (1958)

J.W. Forrester, *Industrial Dynamics* (MIT Press and John Wiley & Sons, Inc, New York, 1961)

J.C. Fransoo, M.J.F. Woters, Measuring the bullwhip effect in the supply chain. Supply. Chain. Manage. **5**, 78–89 (2000)

S. Geary, S.M. Disney, D.R. Towill, (2003) in *Bullwhip in Supply Chains. Past, Present and Future*. 17th International Conference on Production Research, Virginia, USA, p. 3-7 August

J. Holmström, Product range management: a case study of supply chain operations in the European grocery industry. Supply. Chain. Manage. **2**, 107–115 (1997)

J. Holmstrom, K. Framling, R. Kaipia, J. Saranen, Collaborative planning forecasting and replenishment: new solutions needed for mass collaboration. Supply. Chain. Manage. **7**, 136–145 (2002)

J.B. Houlihan, International supply chain management. Int. J. Phys. Distrib. **17**, 51–66 (1987)

S. John, M.M. Naim, D.R., Towill, Dynamic analysis of a WIP compensated decision support system. Int. J.Manuf. Sys. Des. **1**, 283–297 (1994)

H.L. Lee, P. Padmanabhan, S. Whang, The bullwhip effect in supply chains. Sloan. Manage. Rev. **38**, 93–102 (1997a)

H.L. Lee, P. Padmanabhan, S. Whang, Information distortion in a supply chain: the Bullwhip Effect. Manage. Sci. **43**, 543–558 (1997b)

W. Li, P.K. Humphreys, A.C.L. Yeung, E. Cheng, The impact of specific supplier development efforts on buyer competitive advantage: an empirical model. Int. J. Prod. Econ. **106**, 230–247 (2007)

J.D. Sterman, Modeling managerial behavior: misperceptions of feedback in a dynamic decision making experiment. Manage. Sci. **35**, 321–339 (1989)

J.D. Sterman, *Business Dynamics: Systems Thinking and Modeling for a Complex World 353* (McGraw-Hill Higher Education, New York, 2000)

J.M. Swaminathan, S.R. Tayur, Models for supply chains in e-business. Manage. Sci. **49**, 1387–1406 (2003)

D.R. Towill, Dynamic analysis of an inventory and order based production control system. Int. J. Prod. Res. **20**, 369–383 (1982)

D.R. Towill, FORRIDGE. Principles of good practice in material flow. Prod. Plan. Control. **7**, 622–632 (1997)

D.R. Towill, A. Del Vecchio, Application of filter theory to the study of supply chain dynamics. Prod. Plan. Control. **5**, 82–96 (1994)

D.R. Towill, P. McCullen, The impact of an agile manufacturing programme on supply chain dynamics. Int. J. Logist. Manag. **10**, 83–96 (1999)

Y.Z. Tsypkin, *Sampling Systems Theory and Its Application* (Pergamon Press, Oxford, 1964)

P. Wangphanich, S. Kara, B. Kayis, Analysis of the bullwhip effect in multi-product, multi-stage supply chain systems-a simulation approach. Int. J. Prod. Res. **48**, 4501–4517 (2010)

J. Wikner, M.M. Naim, D.R. Towill, The systems simplification approach in understanding the dynamic behaviour of a manufacturing supply chain. Syst. Eng. **2**, 164–178 (1992)

Chapter 4
System Dynamics: Main Concepts

4.1 Introduction

Prior to commencing systems dynamics, it is necessary to know its origin as a discipline or methodology, be it briefly.

The origins of systems dynamics date back to cybernetics which, in turn, originate in Plato and *kibernetes,* André Marie Ampere's government forms (Nineteenth Century), Weiner's study of control and self-control (1948), the art of ensuring the efficacy of action by Couffignal (1966) and closed systems (state of space, variety, entropy). Cybernetics is the interest in the study of the difference between the presence and absence of several properties (dimensions or attributes) (Martín, 2006).

Later, the feed-back concept is introduced into a system. Here are some system definitions:

- Series of elements that are interrelated in such a way that a change in one element affects the whole series Bertalanffy Von (1968).
- Series of variables that the observer selects from those available in the real machine (Ashby 1952).
- Series of parts that work to accomplish a common objective (Forrester 1961).

The general systems theory (Bertalanffy Von 1968) was developed by biologist Ludwig von Bertalanffy in 1940, and is based on the open systems concept (organism systems theory) formed by input variables, output variables and a feedback loop.

In *Industrial Dynamics*, Forrester (1961) presented a methodology for the simulation of dynamic models, which is the origin of systems dynamics (Sterman 2000). Industrial dynamics is a quantitative approach that studies the characteristics of information feedback from industrial systems which are made up of six flows: information, orders, materials, money, personnel and equipment. The four elemental bases of industrial dynamics are the feedback control theory, decision-making processes, the experimental approach and computers development. The first applications centered on production planning and sales.

F. Campuzano and J. Mula, *Supply Chain Simulation*,
DOI: 10.1007/978-0-85729-719-8_4, © Springer-Verlag London Limited 2011

The basic systems dynamics objective is to understand the structural causes that trigger system performance (Martín, 2006). This is a long-term approach. Suitable variables selection is most important (system elements analysis) because it is based on the analysis of internal logic and the system's structural relationships. Forecasting models attempt to supply accurate data about the modeled system's future situation. Management models attempt to establish that alternative *X* is better than *Y*. Systems dynamics devise models of the latter class, while fitting the model to historical data takes second place.

Systems dynamics-based simulation models are mathematical models (abstracts) that are dynamic (interactions that vary with time), linear or nonlinear, stable (tend to return to their initial condition after being disturbed) or unstable, of a stable state (repeated with time) or are transitory (the system's nature is modified with time).

This chapter describes the process to construct a causal loop diagram and a flow diagram, the main systems dynamics tools. It also describes the steps to follow to construct a systems dynamics-based simulation model and revises the most significant tests to validate a dynamic simulation model. Finally, it identifies the level, flow and auxiliary variables to construct a dynamic supply chain model for a traditional, reduced, e-shopping, EPOS and VMI supply chain.

4.2 Causal Loop Diagram

Representing a system with systems dynamics is done by using the causal loop diagram. It includes the key system elements and the relationships among them, based on cause having an influence on effects. Arrows represent relationships and come with a + or a – symbol. The + symbol represents a change in the origin variable of the arrow that will produce a change in the same sense in the destination variable. Then – depicts that the effect produced will take place in the opposite sense. Therefore in a positive relationship, an increase in A causes an increase in B, or a drop in A leads to a drop in B. In a negative relationship however, an increase in A leads to a drop in B, or a drop in A brings about an increase in B.

Feedback represents a closed chain of causal relationships. Loops are positive when the number of negative relationships is even (or zero) and are negative if this number is odd. Negative loops act as stabilizing elements (filling a glass with water, a heating thermostat) and they lead the model toward a stable situation. Positive loops make the system become unstable, irrespectively of the initial situation. The systems containing both loop types and final performance will depend on which one is dominant.

Positive loops trigger systems to grow, evolve and collapse. The most important factor is to understand how the systems' structure produces their performance. The limiting factor is a system whose element limits its growth. It is unique at all times, dynamic and can vary over time.

Key factors are system elements that the system is very sensitive to. One system has several key factors (which are neither evident nor easy to identify). They can be used to bring about important changes to the system with minimum effort, and they tend not to vary over time. Key factors can set off violent performances, and simulation models can identify key factors.

Systems are stable if they are made up of or dominated by a negative loop (an odd number of negative relationships), and are unstable if they are made up of or dominated by a positive loop. A system is hyperstable if formed by many negative loops; in which case, any action that attempts to modify an element is counteracted not only by the loop where this element is located, but by the whole series of negative loops, which overstabilizes the system. A system is sigmoid if a dominant positive loop exists that starts up the system exponentially and if, afterward, a negative loop provides it stability. On the other hand, systems oscillate if they have at least two levels.

An "intelligent" structured typology can be established in systems by centering on the performance they display. Systems' basic structure comprises a desired state and a real state. These two states are compared (difference) and the system carries out an action so that the real state matches the desired one. Generic performance patterns tend to appear irrespectively of the study object:

- One of these patterns is resisting change, with two possible performances:
  - Burnt-out system. A system that is accustomed to receiving many changes, most of which are negative (old system)
  - Idle system. Changes may require initial effort (new system)
- Furthermore, objectives erosion may take place. In this case, performing the action means using resources. The system attempts to avoid expending the energy the action needs, and reconsiders the desired state without having to perform any action. In this way, the system makes no amendment to its real state. There is talk of pollution appearing from the real state to the desired one, and there are two possible performances:
  - The hero system. It tries to convince the system that the effort required by the desired state is of no importance.
  - It obtains an external element that serves as a reference of the desired state so that system pressures cannot amend it; moreover, the system does not have the capacity to amend the external element.
- Addiction to an external element. The external element is a physical element that always has a passive attitude. The effect that the external element has is that the real state matches the desired state; therefore, not any action is necessary. When the external element's effect disappears, the problem arises again.
- Passing the load to the external factor. The system receives help from another external system. In this case, the external system has a will of its own. The problem arises if the system withdraws its help.
- Short-and long-term effects. A contradiction may occur between short-term effects and other long-term ones.

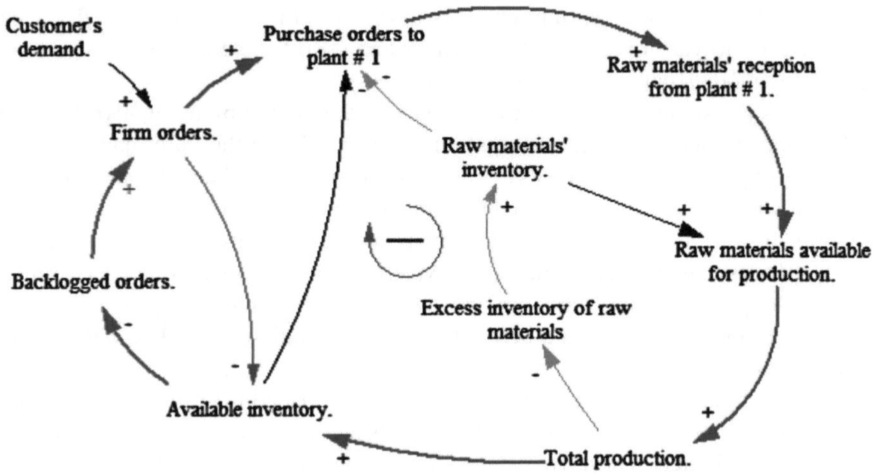

Fig. 4.1 Causal loop for the inventory system in a supply chain

Figure 4.1 shows the causal loop for an inventory system in a supply chain. It is a stable system dominated by a negative loop. For Plant 2, whose raw material supplies are restricted owing to deliveries from Plant 1, the loop controlling both the inventories system and the system launching the production orders starts by consolidating backlogged orders and demand by calculating firm orders. Firm orders are checked against the inventories of finished goods and the raw material available and, from this difference, a purchase order to Plant 1 is generated. Plant 1 delivers the required products with which the material required for the month's production in Plant 2 becomes available. This available material generates total production which increases finished goods inventories, lowers excess raw material inventories, and lowers the number of backlogged orders, thus stabilizing the system.

It is important to highlight the Inventory Order-Based Production Control System (IOBPCS) model developed by Towill (1982) and based on the model proposed by Forrester (1961), and in its evolved model according to other authors (John et al. 1994), APIOBPCS (Automatic Pipeline, Inventory and Order-Based Production Control System). The causal loop diagram for the IOBPCS model is that shown below (Fig. 4.2):

Although the model is a simplified version of real systems, it includes variables like production orders, inventory levels and the lead times common to a large number of companies (Berry 1994). It is also capable of reproducing the performance of these real systems with a high degree of accurateness (Edhill 1990).

The IOBPCS represents a production and inventory control system in which the production level (production order, (*Order Rate ORATE*)) required will depend on the target inventory to be accomplished (*Target Inventory*). To go about this, the production order is based on the average demand (*Average Consumption*) during a certain time period, and also on the actual inventory (*Actual Inventory*). Parameter T_i represents the gain in inventory to be obtained to smooth or amplify the

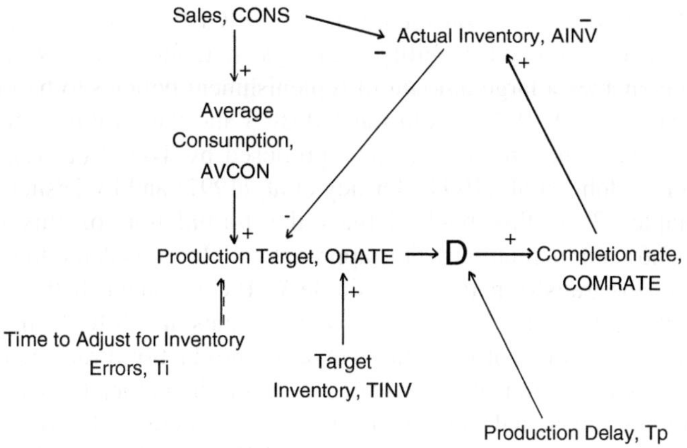

Fig. 4.2 Causal loop diagram of the IOBPCS model. Source: Disney (2001)

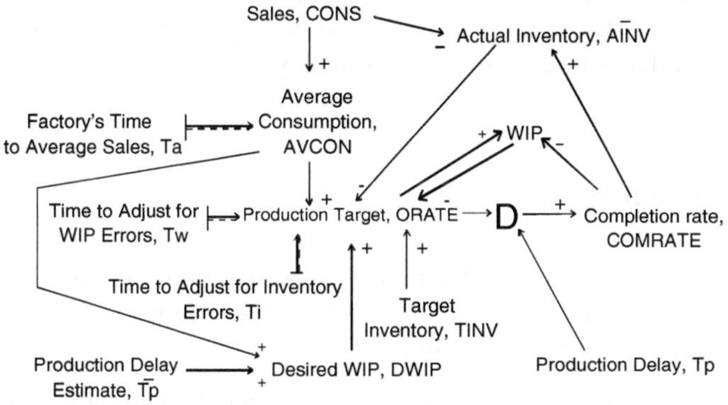

Fig. 4.3 Causal loop diagram of the APIOBPCS model. Source: Disney (2001)

production order, while parameter T_p represents the time needed to manufacture products. The IOBPCS is quite appropriate for moderately constant demand levels (Lewis 1997). However for demands with a high level of variability, the target inventory must be continually updated, which is done in terms of demand.

One characteristic of Forrester's original model (1958), which is not represented in the IOBPCS, is the possibility of there being backlogs as the IOBPCS model is linear and assumes that the whole demand received is served following the corresponding backlog caused by the production of the processed order.

The IOBPCS model evolves to the so-called APIOBPCS (John et al. 1994). Its causal loop diagram is shown in Fig. 4.3.

APIOBPCS can be used to calculate a production order or a production objective via MRP (Material Requirement Planning) systems and may use a list of materials. It may also be used to calculate a purchase order for a supplier if a distributor is involved instead of a manufacturer. As Fig. 4.3 illustrates, the WIP

(Work in Progress) level is considered in the APIOBPCS model to calculate the production order, which adds stability to the system (John et al. 1994). Adding the WIP variable enables a large amount of replenishment policies to be modeled and simulated (Silver et al. 1998). Both the IOPBCS and the APIOBPCS models are constructed around the production order proposed by Towill (1982), which was adapted later by John et al. (1994), Disney et al. (1997) and by Disney and Towill (2001). Chapter 3 of this book defines the formulation of this quantitative expression (3.6) and its relations. This approach will be considered the end product obtained with the transformation of a single WIP component. If the final product had the structure of several components, with amounts that differ from the unit, the previous formula would not be valid. Thus, the production order equals demand forecasting plus a fraction of error which is in the inventory (this is the desired inventory minus the actual net inventory), plus a fraction of error in the work currently underway (WIP) (this is the desired WIP minus the actual WIP). One of the restrictions of the APIOBPCS model is that it does not contemplate the possibility of there being backlogged orders or unsatisfied demand. Backlogged orders are subsequently incorporated into Campuzano's model (2006).

The causal loop diagram is very important to explain the final model to the user and serves as a basis to construct the flow diagram.

4.3 Flow Diagram

The flow diagram, or Forrester diagram, is a translation of the causal loop diagram into a terminology that helps write equations in the computer. The steps to follow for its construction are the following:

- Define the level or state variables, which are a mental photograph of the system and what it results in (warehouse, backlogged orders, etc.). The levels determine the decisions controlling the flow variables.
- Define flow variables, which are the elements determining the variation of levels (products entering and leaving the warehouse, dispatching backlogged orders, etc.).
- Define auxiliary variables, which is the rest of the elements (firm orders, customer demand, etc.)

Fig. 5.6 in Chap. 5 of this book depicts the entry of finished goods in the warehouse which are served in terms of final customer demand. Suppliers send finished goods to the warehouse. Final customer demand of finished goods and backlogged orders comprise the firm order that the warehouse makes up. In this case, the level variables are the warehouse and backlogged orders; the flow variables are the products leaving the warehouse, the products entering the warehouse and the dispatched backlogged orders; the auxiliary variables are firm orders and customer demand. Chapter 5 of this book covers the construction of a simulation model for warehouse management purposes in detail.

Furthermore, the levels are the elements that show the situation the model is in at all times, and a rectangle represents them (see Fig. 5.6). They present an accumulation and vary only in terms of the input and output flows (stock, clearance sale goods, etc.). These levels continue to exist even when the system stops and there are no flows. A level can have any amount of input flow and output flow channels. Some flow diagrams may include levels whose content is inexhaustible, and "clouds" represent them.

Flows cover the actions as a result of the decisions made in the system, by determining the variations of levels. They are temporary functions (stock/month, sale clearance goods/day, etc.) and define the flows present among the system's levels and correspond to activity. Levels determine them in accordance with the rules defined by the decision functions.

Auxiliary and constant variables are parameters that permit a better visualization of the aspects that condition the performance of flows (valves in Fig. 5.6), and are also known as rate equations or decision functions.

The information channels that connect the decision functions with the levels transmit information which, given its nature, is not retained. The materials channels transmit the physical magnitudes between flows and levels.

One other important component of flow diagrams is lags which simulate backlogs in the time taken to transmit materials or information, and act as input variable smoothers. For material backlogs, first-order functions may be used which, when faced with a step input, will respond with an exponentially asymptotic curve. For information lags, a third-order lag may be used which leads to a sigmoid curve.

Having defined the level, flow and auxiliary variables, instructions or equations are then written. There are different software packages, and the most widely used are (in alphabetical order): DYNAMO, ITHINK, POWERSIM, STELA and VENSIM. The following chapters of this book mention that the Vensim® simulation software has been used.

4.4 Constructing a Model

We refer readers to Sterman (2000) and Martín (2006) for a detailed guide to constructing a systems dynamics-based simulation model. This section briefly describes the steps to follow proposed by Martín (2006) and offers an additional step: model validation.

4.4.1 Creating the Causal Loop Diagram

Collecting information about scientific or technical studies that endorse this causal relation or an expert's opinion on the theme to be covered is most useful.

1. Define the problem.
2. Define first-order influences.

3. Define second-order influences. This refers to the name of the elements that have an influence on the first-order elements. The name of these elements must be written around the previous ones.
4. Define third-order influences. The previous process must be repeated with new elements that influence them. Repeat this operation as many times as necessary.
5. Define relations. Draw arrows or the relations among the system elements by assigning a positive or a negative sign to each relation. If the sign of the relation is not clear, it is necessary to redefine the elements.
6. Identify the feedback loops and their signs. Positive loops will be the motors of change, while negative ones will be the causes of system stability. It is necessary to identify the relations where there are backlogged materials or information lags.
7. Refine the non relevant influences. This is a simplification step of the unnecessary elements. The final format must remain as small as possible.
8. Devise possible solutions to the problem. It is necessary to identify (whenever possible) the system performance patterns and to think up solutions for the problem in order to modify the relations among the elements rather than their nature.

4.4.2 Creating the Flow Diagram

The chosen simulation software carries out this phase.

9. Characterize elements. This phase identifies the level, flow and auxiliary variables.
10. Write equations. The relations among the elements are specified through equations. To do this, arithmetic formulae, software functions or tables can be used.
11. Assign values to the parameters. It is necessary to assign a value to the elements, which can be a known one or an approximate one.
12. Create a preliminary version of the model. This is the first model that works, although it can be improved later.
13. Stabilize the model. This consists in the model functioning with all the stable variables.
14. Identify the key elements. The proposals to improve the system will focus on the key elements.
15. Simulate. Amendments to the model can be made which can be applied in the practice.

4.4.3 Model Validation

16. Model validation. After developing the simulation model, it is very important to verify and validate it before simulating different scenarios and making decisions about it. Having verified the model, the researcher ensures that the

constructed model is that which he/she intends to construct; in other words, determine that the simulation model works as expected. This is really a part of model validation. With model validation, the researcher guarantees that the model describes a specific phenomenon; that is, he/she determines the use of the model in relation to a given purpose. Therefore, the objective of model validation is to produce information that helps potential users to accept or reject the model.

Sterman (2000) summarizes the most significant tests to validate a dynamic model.

- Suitability test of the model's limits. Checks if the appropriate concepts have been considered.
- Structural validation test. Validates the model's consistency with the knowledge available about real systems (production and inventory control systems).
- Dimensional consistency test. Checks that the measurement units employed are consistent.
- Test to check similar models. This test reproduces the performance of those models that reproduce similar systems.
- Extreme conditions test. The model performs for real when used under extreme conditions.
- Error integration test. It measures if the model is sensitive to changes in time.
- Reproduction test of known performances. This assesses the model's ability to reproduce the performance of real systems.
- Sensitivity analysis test. This analyzes the model's robustness to face changes in its parameters.
- Anomalous performances test. It assesses (by eliminating or modifying) the existing relationships between the model variables.
- Test of unexpected performance. This tests possible unexpected model performance.
- Test to assess the parameters used. It checks the equivalence of the parameters used and their value with those that really exist.
- Test to assess improvement to the modeled system's operation. It checks if the model can solve the problem for which it has been created.

4.5 Supply Chain Simulation Variables

This section provides the level, flow and auxiliary variables to be considered in the simulation of traditional, reduced, e-shopping, EPOS and VMI supply chains based on the APIOBPCS model of John et al. (1994) and of Campuzano (2006). The remaining chapters of this book consider these variables for supply chain simulation.

The elements used for supply chain simulation will be characterized, and the level, flow and auxiliary variables will be identified. This stage corresponds to Step 9 of the systems dynamics-based simulation model construction (see the previous section), and also to the first step of the flow diagram construction.

The elements to consider in the APIOBPCS-based model for traditional supply chain simulation are:

a. Final customer demand and the demand of a level toward that situated immediately upstream.
b. Firm orders (retailer, wholesaler and manufacturer)
c. Backlogged orders (retailer, wholesaler and manufacturer)
d. On-hand inventory (retailer, wholesaler and manufacturer)
e. Forecasting demand (retailer, wholesaler and manufacturer)
f. Inventory position (retailer, wholesaler and manufacturer)
g. Replenishment orders (retailer and wholesaler)
h. Manufacturing orders (manufacturer)
i. Supply lead times (wholesaler and manufacturer)
j. On-order products (retailer, wholesaler and manufacturer)
k. Manufacturing capacity (manufacturer)
l. Manufacturing (manufacturer)
m. Manufacturing lead times (manufacturer)
n. Fill rate levels (retailer, wholesaler and manufacturer)
o. Inventory costs (storage and order), stockout (retailer, wholesaler and manufacturer)

The elements to consider in the APIOBPCS-based model for reduced supply chain simulation are:

a. Final customer demand and the demand of a level toward that situated immediately upstream.
b. Firm orders (wholesaler and manufacturer)
c. Backlogged orders (wholesaler and manufacturer)
d. On-hand inventory (wholesaler and manufacturer)
e. Forecasting demand (wholesaler and manufacturer)
f. Inventory position (wholesaler and manufacturer)
g. Replenishment orders (wholesaler)
h. Manufacturing orders (manufacturer)
i. Supply time (wholesaler and manufacturer)
j. On-order products (wholesaler and manufacturer)
k. Manufacturing capacity (manufacturer)
l. Manufacturing (manufacturer)
m. Manufacturing lead times (manufacturer)
n. Fill rate levels (wholesaler and manufacturer)
o. Inventory costs (storage and order), stockout (wholesaler and manufacturer)

The elements to consider in an APIOBPCS-based model for e-shopping supply chain simulation are:

a. Final customer demand and the demand of a level toward that situated immediately upstream.
b. Firm orders (manufacturer)
c. Backlogged orders (manufacturer)
d. On-hand inventory (manufacturer)
e. Forecasting demand (manufacturer)
f. Inventory position (manufacturer)
g. Manufacturing orders (manufacturer)
h. On-order products (manufacturer)
i. Manufacturing capacity (manufacturer)
j. Manufacturing (manufacturer)
k. Manufacturing lead times (manufacturer)
l. Fill rate levels (manufacturer)
m. Inventory costs (storage and order), stockout (manufacturer)

The elements to consider in an APIOBPCS-based model for EPOS supply chain simulation are:

a. Final customer demand and the demand of a level toward that situated immediately upstream.
b. Firm orders (retailer, wholesaler and manufacturer)
c. Backlogged orders (retailer, wholesaler and manufacturer)
d. On-hand inventory (retailer, wholesaler and manufacturer)
e. Forecasting demand (retailer, wholesaler and manufacturer)
f. Inventory position (retailer, wholesaler and manufacturer)
g. Replenishment orders (retailer and wholesaler)
h. Manufacturing orders (manufacturer)
i. Supply time (wholesaler and manufacturer)
j. On-order products (retailer, wholesaler and manufacturer)
k. Manufacturing capacity (manufacturer)
l. Manufacturing (manufacturer)
m. Manufacturing lead times (manufacturer)
n. Services levels (retailer, wholesaler and manufacturer)
o. Inventory costs (storage and order), stockout (retailer, wholesaler and manufacturer)

The elements to consider in an APIOBPCS-based model for VMI supply chain simulation are:

a. Final customer demand and the demand of a level toward that situated immediately upstream.
b. Firm orders (wholesaler and manufacturer)
c. Backlogged orders (retailer, wholesaler and manufacturer)
d. On-hand inventory (retailer, wholesaler and manufacturer)

e. Forecasting demand (retailer, wholesaler and manufacturer)
f. Inventory position (retailer, wholesaler and manufacturer)
g. Replenishment orders (retailer and wholesaler)
h. Manufacturing orders (manufacturer)
i. Supply time (wholesaler and manufacturer)
j. On-order products (retailer, wholesaler and manufacturer)
k. Manufacturing capacity (manufacturer)
l. Manufacturing (manufacturer)
m. Manufacturing lead times (manufacturer)
n. Fill rate levels (retailer, wholesaler and manufacturer)
o. Inventory costs (storage and order), stockout (retailer, wholesaler and manufacturer) and penalty costs
p. Maximum inventory level allowed in the retailer's warehouse
q. Minimum inventory level allowed in the retailer's warehouse

4.6 Conclusions

This chapter has reviewed the main systems dynamics concepts by describing the construction process of a causal and a flow diagram, these being the main systems dynamics tools. This chapter has also described the steps to follow to construct a systems dynamics-based simulation model, and has summarized the most significant tests to validate a dynamic simulation model. Finally, and based on the APBIOPCS model, this chapter has identified the level, flow and auxiliary variables required to construct a dynamic model of a traditional, reduced, e-shopping, EPOS and VMI supply chain.

The remaining chapters of this book will practically use the theoretical knowledge covered in this book until this stage.

References

W.R. Ashby, *Design for a Brain* (Wiley, New York, 1952)

L. Von Bertalanffy, *General System Theory: Foundations, Development Applications* (George Braziller, New York, 1968)

D. Berry, The Analysis, *Modelling and Simulation of a Re-engineered PC Supply Chain*. PhD Thesis. University of Wales, Cardiff, (1994)

F. Campuzano (2006) *Model of the Demand Variability Management in the Supply Chain. Analysis of the Bullwhip Effect* (*in Spanish*). PhD Thesis. Universitat Politècnica de València, Departament of Business Management, CEL Price in 2006 to the Best Doctoral Dissertation

L. Couffignal, *La cibernetique* (PUF, Paris, 1966)

S.M. Disney, The production and inventory control problem in vendor managed inventory supply chains. Ph.D. Thesis, Cardiff Business School, Cardiff University, UK, (2001)

S.M. Disney, M.M. Naim, D.R. Towill, Dynamic simulation modelling for lean logistics. Int. J. Phys. Distrib. **27**, 174–196 (1997)

S.M. Disney, D.R. Towill, *The Impact of Vendor Managed Inventory on the Bullwhip Effect in Supply Chains*. Proceedings of the 16th International Conference on Production Research, Prague, (2001)

J. Edhill, *The Application of Aggregate industrial dynamics techniques to manufacturing systems*. PhD Thesis. University of Wales, Cardiff, (1990)

J.W. Forrester, *Industrial Dynamics* (MIT Press and Wiley, New York, 1961)

S. John, M.M. Naim, D.R. Towill, Dynamic analysis of a WIP compensated decision support system. Int. J. Manuf. Syst. Des. **1**, 283–297 (1994)

J. Lewis, *An integrated approach to re-engineering material flow within a seamless supply chain*. PhD Thesis. Cardiff University, Wales, (1997)

J. Martín, *Theory and Practical Exercises of System Dynamics*. ed. by J.M. García, Barcelona, (2006)

E.A. Silver, D.F. Pyke, R. Peterson, *Inventory Management and Production Planning and Scheduling* (Wiley, New York, 1998)

J.D. Sterman, *Business Dynamics: Systems Thinking and Modeling for a Complex World* (McGraw-Hill Higher Education, New York, 2000)

D.R. Towill, Dynamic analysis of an inventory and order based production control system. Int. J. Prod. Res. **20**, 369–383 (1982)

N. Weiner, *Cybernetics or Control and Communication in the Animal and the Machine* (MIT Press, Cambridge, MA, 1948)

Chapter 5
Starting to Model the Supply Chain: Warehouse Management

5.1 Introduction

Before beginning to construct complex models that simulate the demand management process in a multilevel supply chain (each level is one of the companies participating in this chain), a small model representing the input of the finished goods in a warehouse is to be designed, which is basically a subsystem that forms part of the company system in charge of managing the demand that reaches a downstream level of the supply chain which this firm may form part of.

The structure of a system, as presented in previous themes using cause-and-effect diagrams, can appear to be essentially static in nature. However, it assesses system performance (the reader is reminded of the definitions of the key factor and the limiting factor) when faced with external disturbances. It is now necessary to analyze how the endogenous system performance originates. We begin with the different elements that appear in a causal loop diagram, and some of these elements represent variations in terms of the time of the other magnitudes considered in this diagram.

This chapter begins by developing a causal loop diagram and a flow diagram which represent the input of the finished goods in a warehouse. Next, three simulation problems are considered and solved so that the reader becomes familiar with the practical application of the theoretical contents covered in this book to date. Problem 1 proposes developing a flow diagram that represents the input and output of products in a retailer warehouse for the purpose of satisfying customer demand. Problem 2 also considers the management process of possible backlogged orders. Finally, Problem 3 constructs the flow diagram of a company that produces perishable products.

F. Campuzano and J. Mula, *Supply Chain Simulation*,
DOI: 10.1007/978-0-85729-719-8_5,

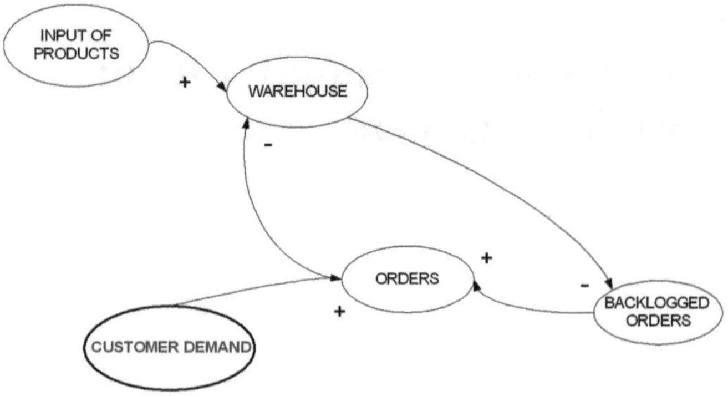

Fig. 5.1 Causal loop diagram of a *warehouse*

5.2 Nature of a Cause-and-Effect Diagram's Dynamic Performance: A Case of a Products Warehouse Management

In accordance with this book's subject matter, and as mentioned earlier, this section considers the specific case of a causal loop diagram which represents the input of products in a warehouse, which are delivered in accordance with the final customer's demand (Fig. 5.1).

The description of this products input and output process with a warehouse can be broken down as follows:

- Finished products demand along with possible existing backlogged orders make up a firm order to be placed. An increase or drop in demand and/or backlogged orders will be reflected in the size of the firm order (Fig. 5.2).
- The more firm orders there are, the lesser the warehouse's capacity will be (Inventory) to respond to demand (Fig. 5.3).
- The more the input of products arriving at the warehouse from the supplier, the higher the available inventory in this warehouse will be.
- The lower the available inventory in the warehouse, the more probabilities there will be of a firm order not being completed; then backlogged orders will increase.

After presenting the different influences among all the system variables represented in the diagram in Fig. 5.1, the system's endogenous operation will be explained.

Among the various elements appearing in the nodes of the former causal loop diagram, some represent variations in relation to the time corresponding to the other magnitudes considered in this same diagram. For instance, in the diagram in Fig. 5.4, the *Input of products* variable represents variation in relation to the warehouse's inventory level time. Figure 5.5

This influence is a particular case of another more general one, which we can express as below:

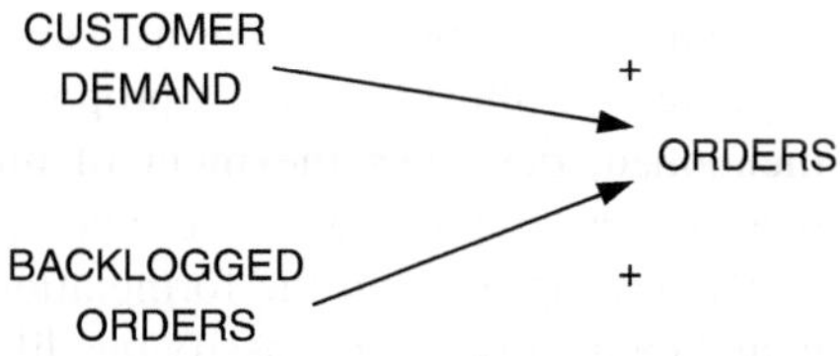

Fig. 5.2 Causal loop *Backlogged orders–Customer demand-Orders*

Fig. 5.3 Causal loop *Orders-Inventory*

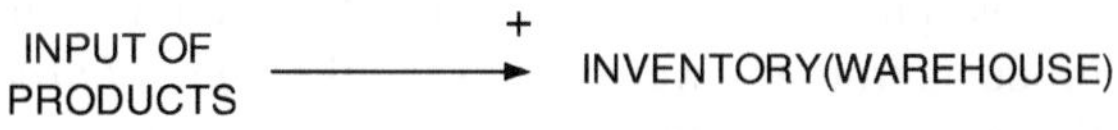

Fig. 5.4 Causal loop *Input of products-Inventory*

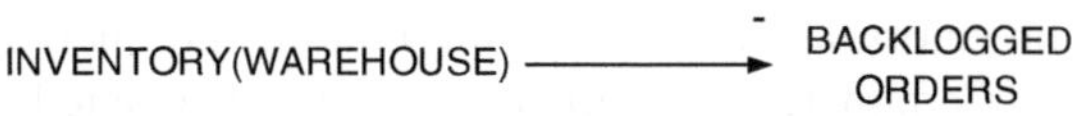

Fig. 5.5 Causal loop *Inventory-Backlogged orders*

$$\frac{\delta X}{\delta t} \rightarrow X \tag{5.1}$$

where $\frac{\delta X}{\delta t}$ denotes variation in relation to magnitude X. This expression represents a trivial relationship: variation in relation to the time of X influences the growth of variable X itself. Nonetheless, the interesting point to stress is that the causal loop diagram contains variables representing variation in relation to the time of the other variables involves the latter varying over time. This fact allows us to state that the system performance in the structure is implicit.

It is also worth observing that whenever there is a variable of type $\frac{\delta X}{\delta t}$, which represents the variation of magnitude X in relation to time, there will be a relationship with influence as in (5.1). Variable X results from the accumulation of the implicit change in variable $\frac{\delta X}{\delta t}$. Thus, whenever a variable appears such as $\frac{\delta X}{\delta t}$, an X will appear and, between them both, a relationship will be established like that described in (5.1). Variable X is known as the *level variable* and variable $\frac{\delta X}{\delta t}$ is called the *flow variable*. In the mathematical literature, the *level variable* is also known as the state variable.

These previous considerations lead to classify (Forrester 1961) the different variables in the diagram of influences into three groups:

- *State or level variables*
- *Flow variables and*
- *Auxiliary variables.*

Level variables are normally the most important and represent magnitudes whose evolution is particularly significant. There are one or several flow variables in association with each level variable, which determine their variation over time. Finally, auxiliary variables are the remaining variables which appear in the diagram. They represent intermediate steps to determine flow variables using level variables. To illustrate this, a propose example will be used which, as previously mentioned, describes the input of finished products in a warehouse which are served in accordance with the final customer's demand.

By bearing in mind the formulations corresponding to the case under study, it would be interesting to classify the different elements that appear in the diagram of Fig. 5.1 into the three proposed variable types: level, flow and auxiliary. To go about this, the level variables of the process in question must be firstly identified. In this case, the level variables will be *Warehouse* and *Backlogged orders*. The *Products delivered* variable (which was not required in the causal loop diagram as it was quite clear that any order that could be met would involve an output of products from the warehouse) and the *Input of products* variable are flow variables that represent, on the one hand, a drop in the warehouse's inventory level when firm orders are delivered and, on the other, an increase in the warehouse's inventory level thanks to the input of products from the supplier.

Certain icons (graphic ones) are associated with level and flow variables. A rectangle is associated with a level variable and a valve-like icon is related with the flow variable, whose opening is regulated precisely by means of the flow that this variable represents. Auxiliary variables tend to be represented by circles, but this is not necessary. If we classify the causal loop diagram‘s components into level, flow or auxiliary variables, and if we associate these elements with their corresponding icons, the *Forrester's diagram* (Forrester 1961) or a flow-level diagram is obtained. In this diagram, the different relationships among the variables they affect are represented by arrows.

The physical magnitudes between flows and levels are transmitted through the so-called "material channels".

In general terms, not only flow of materials in systems, but also information; flow through information channels, and can be used to obtain variables (from information about historical demand values, the demand forecasting auxiliary variable may be obtained); having used the information obtained through the information channel, this can stop flowing through the system.

Finally, all that remains is to define "delays", which represent the time delays in transmitting materials or information. In short, this is the consumption of the *time* resource in the transformation, delivery of materials and transmission of information. In socio-economic systems, there are frequently delays in transmitting information and materials, which is considerably important for system performance.

Figure 5.6 depicts the Forrester's diagram corresponding to the causal loop diagram of Fig. 5.1 (note the distinction made for the various variables used).

The *Warehouse* and *Backlogged orders* variables have been identified as level variables, while *Input of products* and *Products delivered* are flow variables. The *Products delivered* variable represents the variation of the level of the *Backlogged orders* and *Warehouse* variables. However, to provide a better understanding of the Forrester's diagram, the backlogged orders variation has been associated with a new flow variable, *Backlogged orders delivered* which, as seen in the diagram,

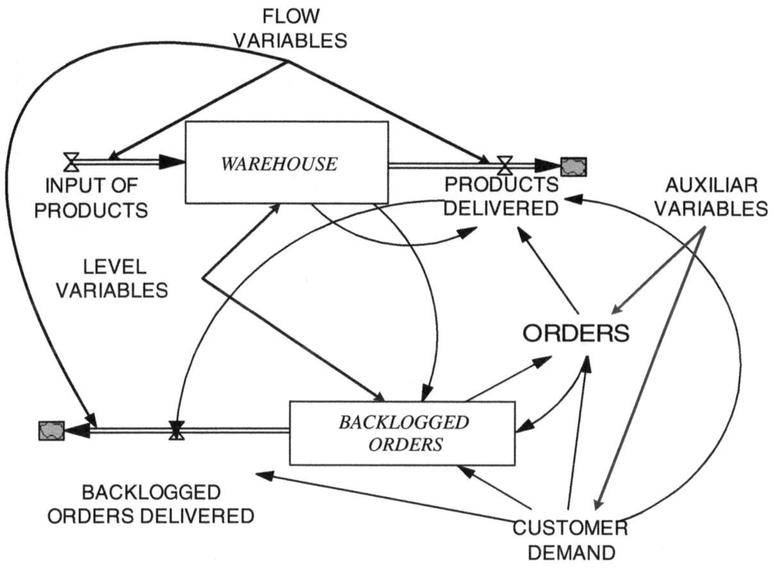

Fig. 5.6 Forrester's diagram representing an input and output system of finished products in a *warehouse*

relates with the *Products delivered* variable**.** The former will depend on the latter's operation, thus the concept defined in the original causal loop diagram will not vary. After identifying the variables, the auxiliary variables, *Orders* and *Customer demand,* are an intermediate step in determining the output of products by using the *Warehouse* and *Backlogged orders* variables.

Finally, and also in Fig. 5.6, a cloud appears that represents a source, or a drain; it is neither relevant for the system's description nor essential, but is included to make the diagram more coherent.

5.3 Practice Problems

The objective of the exercises in this chapter is to help the reader recognize the variables required to model a system that represents how a small retailer manages customer demand by controlling its inventories level and placing orders to a supplier or manufacturer whenever necessary. The solution to these problems serves merely as a guideline and intends to guide the reader to create systems that manage inventories.

The Vensim® program is used to do these exercises (www.vensim.com). We choose this software because it uses a modeling approach that combines systems dynamics concepts (Forrester 1961; Sterman 2000; Amir 2005) and the simulation of discrete events to represent a supply chain's events and uncertainties in detail, and to subsequently analyze its performance with its structure and any existing

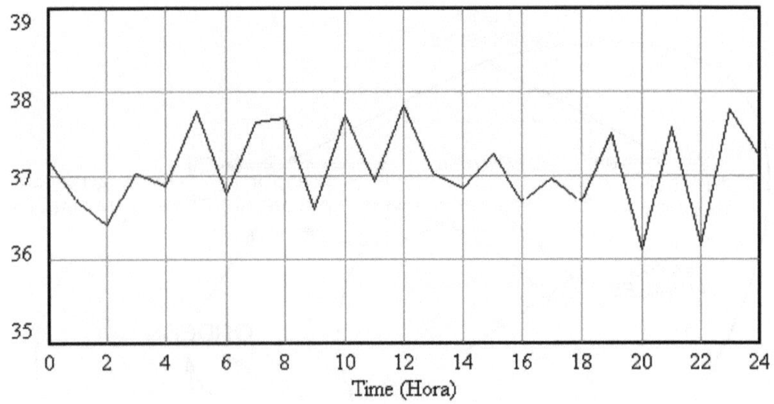

Fig. 5.7 Example of the *Random Uniform* function

causal relations among its components. Local simulation is used to model the proposed problem examples.

5.3.1 Problem 1

Using Vensim®, draw the flow diagram that represents the input and output of products in a retailer warehouse for the purpose of meeting customer demand (*in this case, do not include backlogged orders management; that is, all the customer order not delivered on time due to insufficient available stock* (*stockout*) *is lost*). To meet customer demand, orders will be placed with a manufacturer (in lots of 50 units). The manufacturer's warehouse is assumed to have an unlimited available stock of products. The retailer warehouse has an initial stock of 50 units and the retailer's customer demand follows a uniform distribution with a minimum of 3 units and a maximum of 6 units per period. The manufacturing lead time to the retailer warehouse is 2 periods. Simulate 100-periods.

Observations: to correctly solve this exercise, functions *Random Uniform* to simulate demand and *Delay Fixed* to simulate deliveries to the warehouse after the manufacturing lead time must be used.

RANDOM UNIFORM (*m*, *x*, *s*)
It provides a series of random values with a minimum of *m* and a maximum of *x*. *s* is the calculation parameter of the random numbers, and can take any figure. If *s* changes, the series of random numbers also changes.

Example: the intention is to simulate the performance of a body temperature which ranges between 36 and 38 degrees during the daytime.

Temperature = **RANDOM UNIFORM (36, 38, 99)**

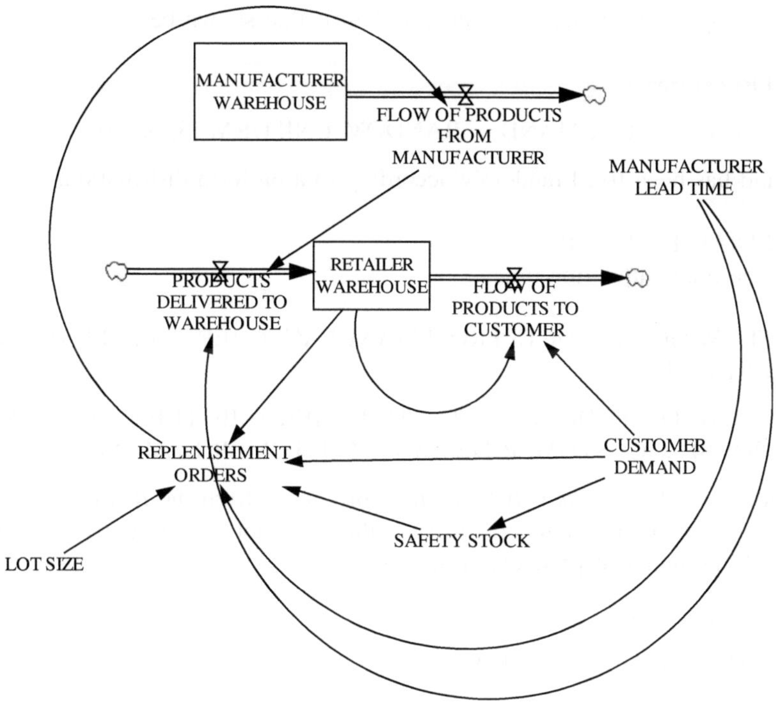

Fig. 5.8 Flow diagram–Problem 1

See Fig. 5.7

DELAY FIXED (*X*, *T*, *N*)

Delay in the step for value *X* and period *T*, starting the simulation with the *N* value instead of the *X* value. *N* tends to be 0.

5.3.2 Solution to Problem 1

The proposed flow diagram corresponding to the system could be as follows:

A variable is considered, *Retailer Warehouse,* which uses a Manufacturer (which, for simplicity reasons, is assumed to have a very large available stock of products so that it is not necessary to model a manufacturing process) to meet the retailer's products demand. The replenishment order is formulated in terms of the variables *Customer demand*, *Manufacturing lead time* (that is, the time it takes the *Manufacturer* to receive an *Order* and for this to reach the *Retailer warehouse*), and of a certain *Safety stock* used to cover the retailer's demand during the manufacturing lead time, thus avoiding stockouts. The *Replenishment orders* from the *Retailer warehouse* to the *Manufacturer* are prepared in lots of 50 units. Figure 5.8

The proposal formulated for this problem is that shown below[1]:

(01) PROBLEM 1

(02) CUSTOMER DEMAND = RANDOM UNIFORM (3, 6, 99)

Demand has been fixed randomly according to a uniform distribution.

(03) FINAL TIME = 100
The final simulation time.

(04) FLOW OF PRODUCTS FROM MANUFACTURER = REPLENISHMENT ORDERS

(05) FLOW OF PRODUCTS TO CUSTOMER = IF THEN ELSE (WAREHOUSE > CUSTOMER DEMAND, CUSTOMER DEMAND, 0)

The *Flow of products* variable, which originates from the *Retailer warehouse* variable, must be programmed so that if there are not enough units to meet total demand, the outlet of products must be 0.

(06) INITIAL TIME = 1
The initial simulation time.

(07) LOT SIZE = 50

(08) MANUFACTURER LEAD TIME = 2

(09) MANUFACTURER WAREHOUSE = -FLOW OF PRODUCTS FROM MANUFACTURER,
Initial Value: 500,000

The *Manufacturer's* initial inventory is very large to ensure no stockouts during the fixed simulation period,

(10) PRODUCTS DELIVERED TO WAREHOUSE = DELAY FIXED (FLOW OF PRODUCTS FROM MANUFACTURER, MANUFACTURER LEAD TIME, 0)

The *Products delivered to warehouse* variable has been fixed with the *DELAY FIXED* variable that stores the products delivered from the *Manufacturer warehouse* and permits the input of these products in our Warehouse, but only after the fixed lead time.

---

1 (*Note*: units have been omitted. The reader can use the units he/she wants).

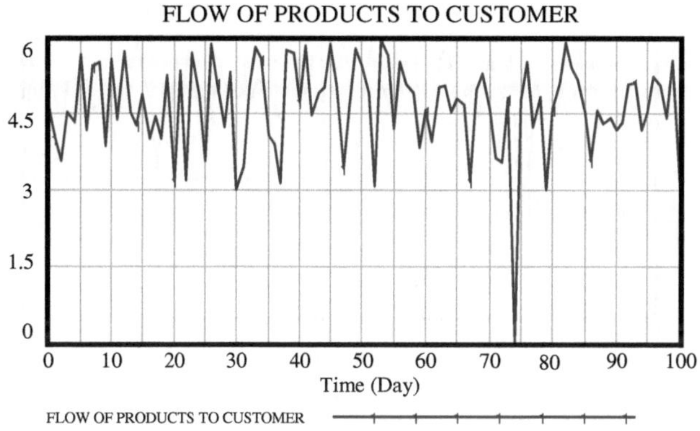

Fig. 5.9 *Flow of products to customer*

(11) REPLENISHMENT ORDERS = IF THEN ELSE (WAREHOUSE < CUSTOMER DEMAND * MANUFACTURER LEAD TIME + SECURITY STOCK * MANUFACTURER LEAD TIME, LOT SIZE, 0)

The *Replenishment order* formulation is very simple. If the number of units available in the *Retailer warehouse* is lower than *Customer demand* multiplied by the Manufacturing *lead time,* plus the *Safety stock* multiplied by this lead time, then the *Replenishment order* is launched to the *Manufacturer.*

(12) RETAILER WAREHOUSE = PRODUCTS DELIVERED TO WAREHOUSE – FLOW OF PRODUCTS TO CUSTOMER
Initial Value: 50

The initial inventory is 50 units

(13) SAVEPER = TIME STEP

(14) SAFETY STOCK = CUSTOMER DEMAND

(15) TIME STEP = 1

The time step for the simulation.

After drawing and simulating the model with Vensim®, the reader should obtain the following graph that corresponds to the *Flow of products to customer* variable (see Fig. 5.9).

The reader can see how the *Flow of products to customer* variable is 0 on period 74. This figure coincides with a stockout period corresponding to our *Warehouse*. How would the reader solve this stockout period? One solution (the reader should analyze other solutions such as modifying the *Replenishment order* variable) would be, for example, to increase the lot size to 60 units. Thus the *Flow of products to customer* would end up as follows (see Fig. 5.10):

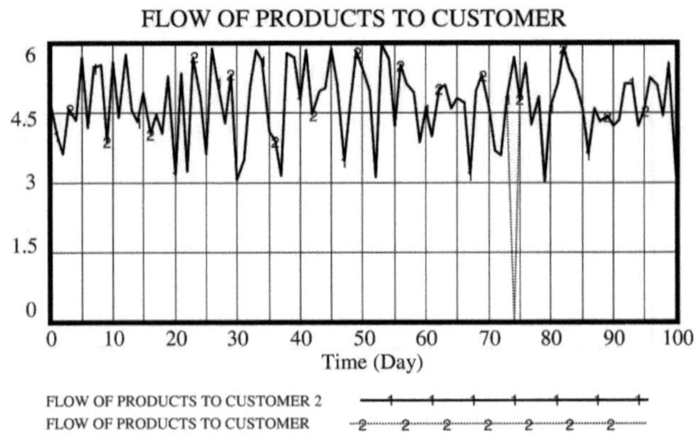

Fig. 5.10 New simulation for *Flow of products to customer*

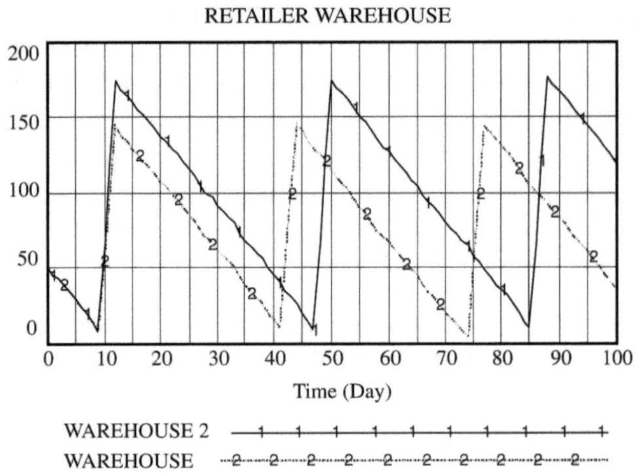

Fig. 5.11 New simulation for *Retailer warehouse*

Observe how the graphs coincide, except during the stockout period as this period has disappeared in the new simulation (SUBINDEX 2).

This change can also be seen in the remaining model variables. For example, *Retailer warehouse* achieves a different amount of units in each period, which is superior to the values obtained when the lot size was 50 units. This will increase the retailer's inventory costs (see Fig. 5.11).

How would the reader solve the problem of increased inventory costs and stockout without having to increase the lot size to 60 units?

Observation: check the *Replenishment order* USED and *Flow of products to customer.* Should a relationship exist between *Products delivered to warehouse* and *Flow of products to customer*?

5.3.3 Problem 2

Using Vensim®, draw the flow diagram that represents the input and output of products to the retailer warehouse, *which includes the management process of possible backlogged orders*. This means that when an order cannot be met, it is delivered when the warehouse has sufficient available stock to do so. In order to meet customer demand, orders will be placed with the manufacturer (in lots of 50 units). Bear in mind that the Manufacturer Warehouse has unlimited available stock. The retailer warehouse has an initial stock of 50 units and the retailer customer demand follows a uniform distribution with a minimum of 3 units and a maximum of 6 units per period. The manufacturing lead time to the retailer warehouse is 2 periods. Simulate 100-periods. How would the reader measure the *Fill Rate* achieved by the retailer?

OBSERVATIONS: The fill rate measures the percentage of delivered orders in relation to the total number of orders placed by the final customer. These are level variables as they vary with time depending on whether or not different orders are delivered. Use the XIDZ function for the formulation.

XIDZ (A, B, X)
The result is *A*/*B*, except when $B = 0$, whose result is *X*. We use this when we have to do the *A*/*B* division; *B* may be zero at some point, which results in the quotient having an infinite value, thus making the model unfeasible. In this case, if *B* is equal to zero, the quotient result is *X*.

5.3.4 Solution to Problem 2

This problem is expressed as the previous one, except for the fact that three new variables are introduced, which enable the management of possible delayed orders not being delivered on time (backlogged orders). These three variables are *Orders*, *Backlogged orders* and *Backlogged orders delivered.* At all times, *Orders* will be possible demand at a given time, plus the *Backlogged orders* to be delivered. When the *Retailer warehouse* has the amount of products available that the *Orders* variable marks, a *Flow of products to customer* will come about. At the same time, the *Backlogged orders delivered* variable will eliminate the amount of these *Backlogged orders delivered* to the customer from the *Backlogged orders* variable. The other two variables to be added are the *Fill rate* and the *Number of simulations*. Now we move on to the model formulation[2]: Fig. 5.12

[2] (*Note*: units have been omitted. The reader can use the units he/she wants).

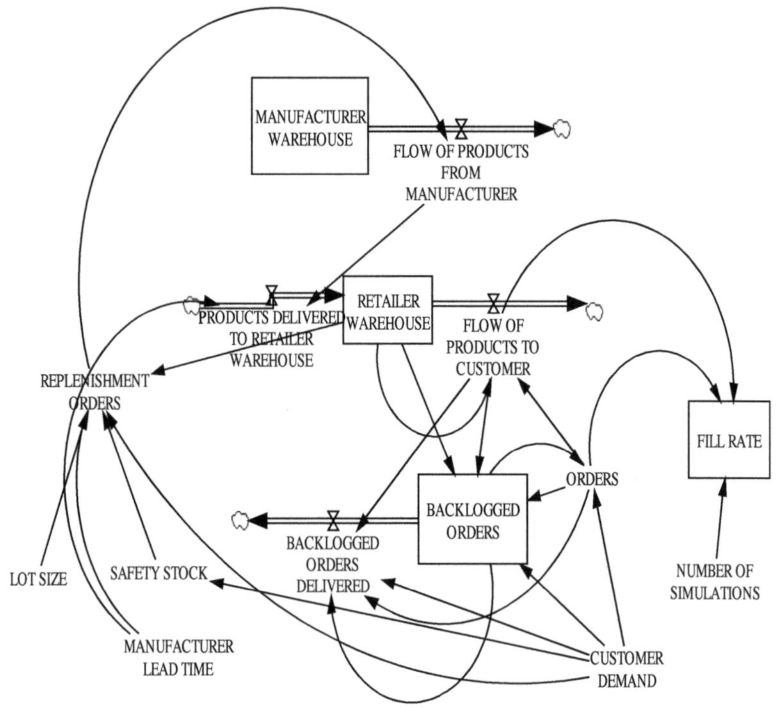

Fig. 5.12 Flow diagram–Problem 2

(01) PROBLEM 2

(02) BACKLOGGED ORDERS = IF THEN ELSE(WAREHOUSE-ORDERS >=0, 0, IF THEN ELSE (FLOW OF PRODUCTS TO CUSTOMER < CUSTOMER DEMAND, CUSTOMER DEMAND−FLOW OF PRODUCTS TO CUSTOMER, 0))−BACKLOGGED ORDERS DELIVERED

Initial Value: 0

(03) BACKLOGGED ORDERS DELIVERED = IF THEN ELSE(FLOW OF PRODUCTS TO CUSTOMER = ORDERS, BACKLOGGED ORDERS, IF THEN ELSE(FLOW OF PRODUCTS TO CUSTOMER > CUSTOMER DEMAND, FLOW OF PRODUCTS TO CUSTOMER−CUSTOMER DEMAND)

Initial Value: 0

(04) CUSTOMER DEMAND = PULSE(1, 2) * 0 + PULSE(2,100) * RANDOM UNIFORM(3, 6, 99)

To simulate demand, the PULSE function has been used to avoid errors at the beginning of the simulation since several variables use the demand variable at the same time, which generates warnings. The error is similar to the circular references in Microsoft Excel® spreadsheet software. For further information, consult the guide in Vensim®.

(05) FILL RATE = XIDZ(FLOW OF PRODUCTS TO CUSTOMER, ORDERS, 1)/(NUMBER OF SIMULATIONS) * 100, Initial Value: XIDZ (FLOW OF PRODUCTS TO CUSTOMER, ORDERS, 1)/(NUMBER OF SIMULATIONS) * 100)

The XIDZ function avoids a program error when the *Orders* variable is 0. In this case, the fill rate at this time is assigned a value of 1 as it is considered that 100% of the customer's requirements are fulfilled in this period, even though there are no firm orders.

(06) FINAL TIME = 100
The final simulation time.

(07) FLOW OF PRODUCTS FROM MANUFACTURER = REPLENISHMENT ORDERS

(08) FLOW OF PRODUCTS TO CUSTOMER = IF THEN ELSE((WAREHOUSE−ORDERS) >= 0, ORDERS, 0)

(09) INITIAL TIME = 1
The initial simulation time.

(10) LOT SIZE = 50

(11) MANUFACTURER LEAD TIME = 2

(12) MANUFACTURER WAREHOUSE = −FLOW OF PRODUCTS FROM MANUFACTURER,
Initial Value: 500,000

(13) NUMBER OF SIMULATIONS = 100

(14) ORDERS = CUSTOMER DEMAND + BACKLOGGED ORDERS

(15) PRODUCTS DELIVERED TO WAREHOUSE = DELAY FIXED(FLOW OF PRODUCTS FROM MANUFACTURER, MANUFACTURER LEAD TIME, 0)

(16) REPLENISHMENT ORDERS = IF THEN ELSE(WAREHOUSE < CUSTOMER DEMAND * MANUFACTURER LEAD TIME + SECURITY STOCK*MANUFACTURER LEAD TIME, LOT SIZE, 0)

(17) RETAILER WAREHOUSE = PRODUCTS DELIVERED TO WAREHOUSE−FLOW OF PRODUCTS TO CUSTOMER
Initial Value: 10
This time the initial *Warehouse* value programmed is 10 units in order to quickly force the appearance of *Backlogged orders.*

(18) SAVEPER = TIME STEP

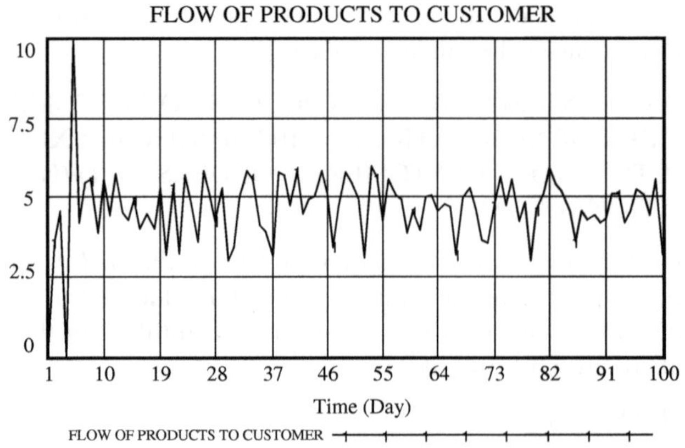

Fig. 5.13 Simulation for *Flow of products to customer*

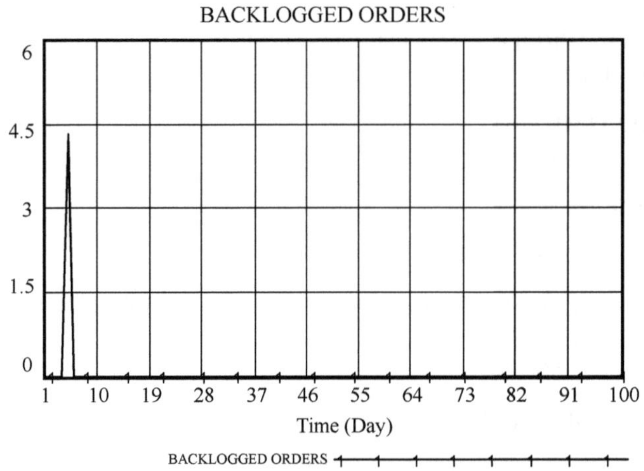

Fig. 5.14 Simulation for *Backlogged orders*

(19) SECURITY STOCK = CUSTOMER DEMAND

(20) TIME STEP = 1

The time step for the simulation.

After drawing and simulating the model in Vensim®, the reader should obtain the following graph corresponding to the *Flow of products to customer* (see Fig. 5.13)

Observe that there is a period with *Backlogged orders* that the model must solve by delivering this amount at the time the *Retailer warehouse* has sufficient units available. The figures below depict this process (see Figs. 5.14 and 5.15).

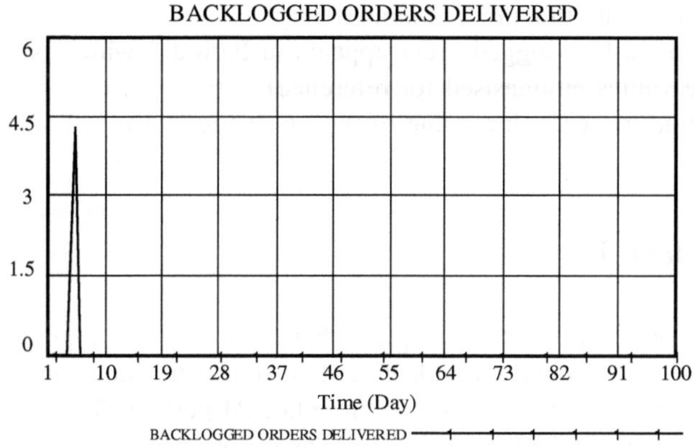

Fig. 5.15 Simulation for *Backlogged orders delivered*

Table 5.1 Mechanism to solve the simulated system's backlogged orders

Time (Day)	Customer demand	Orders	Warehouse	Flow of products to customer	Baclogged orders	Backlogged orders delivered
1	0	0	10	0	0	0
2	3.60426068	3.604260683	10	3.604260683	0	0
3	4.55460787	4.554607868	6.395739555	4.554607868	0	0
4	4.32566118	*4.325661182*	1.841131687	0	0	0
5	5.67177105	9.997432709	51.84113312	9.997432709	*4.325661182*	*4.325661182*
6	4.17777205	4.177772045	91.84370422	4.177772045	0	0
7	5.4533906	5.453390598	137.6659241	5.453390598	0	0

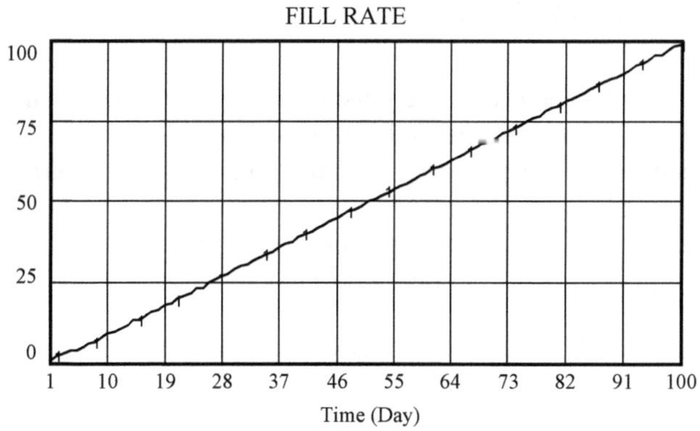

Fig. 5.16 Simulation for the *Fill Rate*

Table 5.1 presents this process numerically:

Observe how a backlogged order appears in Period 4 which is rapidly met in Period 5 (see values emphasised for reference).

The *Fill rate* must reach a value of 99% (see Fig. 5.16).

5.3.5 Problem 3

Using Vensim®, draw the flow diagram that represents a company which manufactures perishable products (evidently, it includes the input and output of the perishable products in the warehouse). Simulate 24-periods. The characteristics of this company's manufacturing process are as follows:

- The warehouse is assumed to have an unlimited maximum manufacturing capacity. The company's manufacturing capacity ranges between 5 and 7 tons of products per period to fulfill demand which ranges between 3 and 10 tons per period (both follow a uniform distribution). This fluctuation in manufacturing depends on the plant's manufacturing capacity each period, which may vary owing to maintenance work, breakdowns, etc. However, the manufacturing department programs it in accordance with the studies done about different demand pattern records. The manufacturing lead time for each manufacturing order is 2 periods.
- Once the warehouse has acquired 10 tons of stored products, if this amount is maintained for more than two periods running because of a drop in sales, it has been verified that 25% of the manufactured articles spoil and have to be withdrawn. The warehouse has an initial stock of 20 tons. What fill rate will the company reach?
- *Observations*: Review the use of the RANDOM UNIFORM and DELAY FIXED formulae.

5.3.6 Solution to Problem 3

This exercise includes the use of counters to manage the time in which the amount of products in the *Warehouse* exceeds 10 tons.

In accordance with the proposed formulation, the flow diagram could be as follows:

The problem formulation will be as follows[3]: Fig. 5.17

(01) EXERCISE 3

(02) COUNTER = INTEG (ONE-COUNTER TO 0, 0)

[3] (*Note*: units have been omitted. The reader can use the units he/she wants apart from those considered in the model formulation).

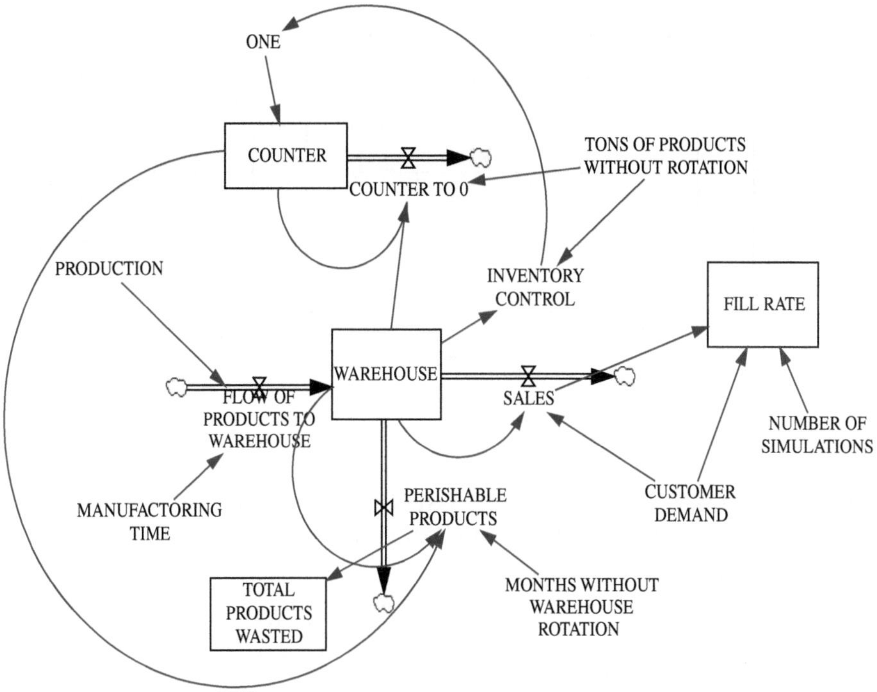

Fig. 5.17 Flow diagram–Problem 3

(03) COUNTER TO 0 = IF THEN ELSE (WAREHOUSE < TONS OF PRODUCTS WITHOUT ROTATION, COUNTER, 0)

This variable places the counter at 0 when the *Warehouse* has a smaller amount than 10 tons available.

(04) CUSTOMER DEMAND = RANDOM UNIFORM (3, 10, 99)

(05) FILL RATE = (XIDZ (SALES, CUSTOMER DEMAND, 1)/NUMBER OF SIMULATIONS) * 100
Initial value:
FILL RATE = (XIDZ (SALES, CUSTOMER DEMAND, 1)/NUMBER OF SIMULATIONS) * 100

(06) FINAL TIME = 24
The final simulation time.

(07) FLOW OF PRODUCTS TO WAREHOUSE = DELAY FIXED (PRODUCTION, MANUFACTURING TIME, 0)

(08) INITIAL TIME = 0
The initial simulation time.

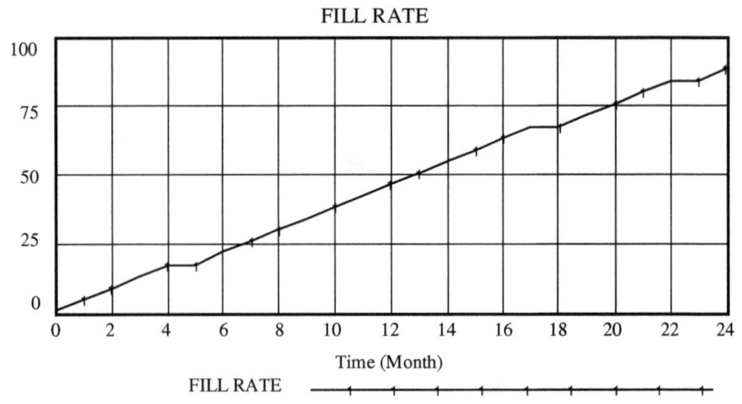

Fig. 5.18 Simulation for *Fill Rate*

(09) INVENTORY CONTROL = IF THEN ELSE(WAREHOUSE > TONS OF PRODUCTS WITHOUT ROTATION, 1, 0)

(10) MANUFACTURING TIME = 2

(11) MONTHS WITHOUT WAREHOUSE ROTATION = 2

(12) NUMBER OF SIMULATIONS = 24

(13) ONE = INVENTORY CONTROL

(14) PERISHABLEPRODUCTS = IF THEN ELSE (COUNTER > MONTHS WITHOUT WAREHOUSE ROTATION, WAREHOUSE * 0.25, 0)

(15) PRODUCTION = RANDOM UNIFORM (5, 7, 99)

(16) SALES = IF THEN ELSE(WAREHOUSE >= CUSTOMER DEMAND, CUSTOMER DEMAND, 0)

(17) SAVEPER = TIME STEP

(18) TIME STEP = 1
The time step for the simulation.

(19) TONS OF PRODUCTS WITHOUT ROTATION = 10

(20) TOTAL PRODUCTS WASTED = PERISHABLE PRODUCTS

Initial Value: 0

This variable indicates the total products that have been wasted during the simulated period.

(21) WAREHOUSE = FLOW OF PRODUCTS TO WAREHOUSE−PERISHABLE PRODUCTS−SALES
Initial Value: 20

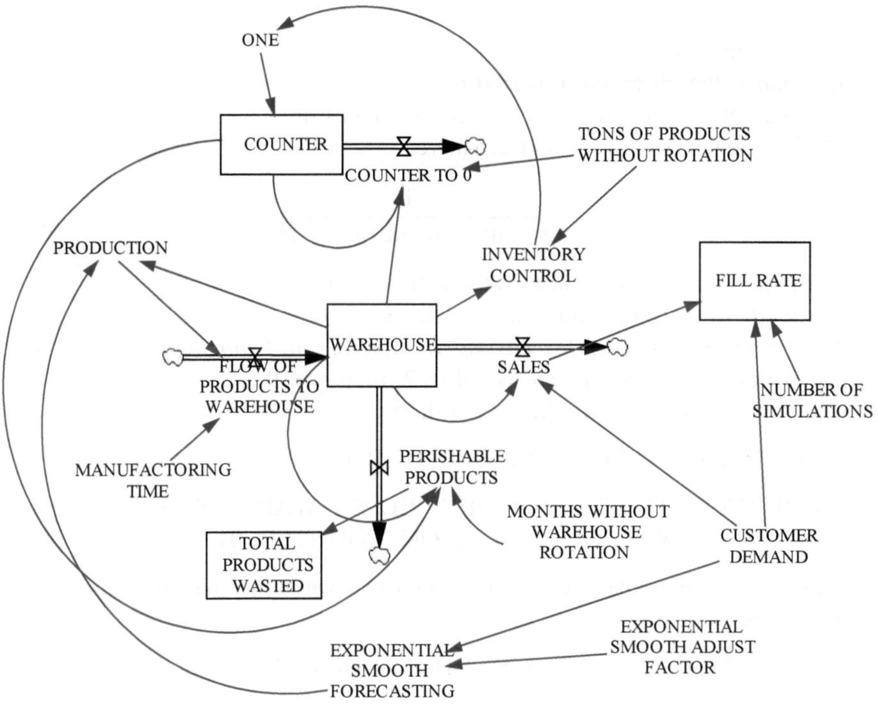

Fig. 5.19 Flow diagram for the new variables

The *Fill Rate* reached will be 84.33% during this 24-period (see Fig. 5.18).

How can you manage to increase the *Fill Rate* without increasing the inventory level in excess and, therefore, the percentage of *Wasted Products*?

One solution guide to this problem can involve modifying the manufacturing order. Between 5 and 7 tons are manufactured every period. As mentioned previously, this fluctuation depends on the manufacturing capacity each period, which may vary due to maintenance work, breakdowns, etc. If the manufacturing order is modified, it is possible to increase the fill rate without increasing the amount in wasted tons in excess over the 24-period period. One way of managing this is by using sales forecastings. *Production* could be amended depending on the *Warehouse* state and on the Sales forecastings. New variables are included in the model for the purpose of checking what will happen. The new flow diagram is that shown below (Fig. 5.19):

Forecastings will be made using the SMOOTH function, which corresponds with an *Exponential Smooth* so that:

The formula employed for forecastings using Exponential Smooth is:

$$Y_{t+1} = Y_t + \alpha \cdot (x_t - Y_t) \tag{5.2}$$

where:

Y_{t+1} = forecasting for period t + 1

x_t = real value observed in period t

α = weight assigned to the most recent forecasting.

The α value is between 0 and 1, where:

$$\alpha = \frac{1}{\text{exponential smooth adjust factor}} \tag{5.3}$$

If α tends to be 1, the adjust factor will be substantial.

If α tends to be 0, the adjust factor will be weak.

The *Exponential smooth forecasting* variable has an *Exponential smooth adjust factor,* which will be considered equal to 2, then $\alpha = 0.5$.

Thus, the formulation would be as follows:

(01) COUNTER = INTEG (ONE-COUNTER TO 0, 0)

(02) COUNTER TO 0 = IF THEN ELSE (WAREHOUSE < TONS OF PRODUCTS WITHOUT ROTATION, COUNTER, 0)

This variable puts the counter at 0 when the warehouse has an amount lower than 10 tons available.

(03) CUSTOMER DEMAND = RANDOM UNIFORM (3, 10, 99)

(04) EXPONENTIAL SMOOTH ADJUST FACTOR = 2

(05) EXPONENTIAL SMOOTH FORECASTING = SMOOTH(CUSTOMER DEMAND, EXPONENTIAL SMOOTH ADJUST FACTOR)

(06) FILL RATE = (XIDZ (SALES, CUSTOMER DEMAND, 1)/NUMBER OF SIMULATIONS) * 100

Initial value: FILL RATE = (XIDZ (SALES, CUSTOMER DEMAND, 1)/ NUMBER OF SIMULATIONS) * 100

(07) FINAL TIME = 24
The final simulation time.

(08) FLOW OF PRODUCTS TO WAREHOUSE = DELAY FIXED(PRODUCTION, MANUFACTURING TIME, 0)

(09) INITIAL TIME = 0
The initial simulation time.

(10) INVENTORY CONTROL = IF THEN ELSE (WAREHOUSE > TONS OF PRODUCTS WITHOUT ROTATION, 1, 0)

(11) MANUFACTURING TIME = 2

(12) MONTHS WITHOUT WAREHOUSE ROTATION = 2

(13) NUMBER OF SIMULATIONS = 24

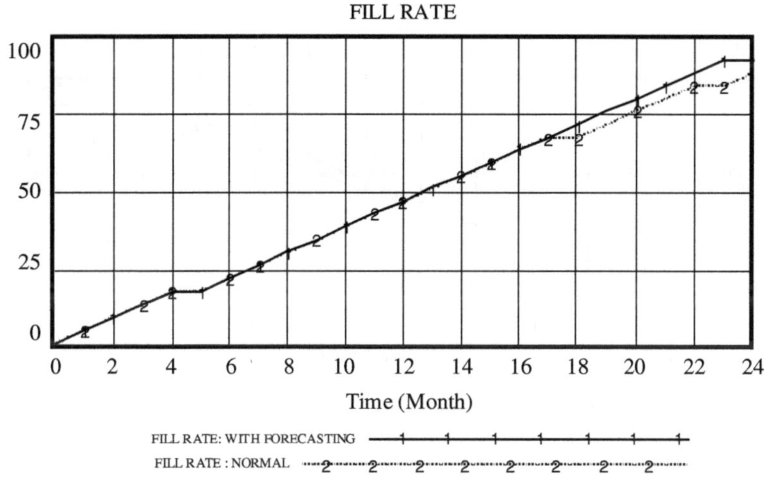

Fig. 5.20 Simulations for *Fill Rates* (different types of replenishment orders)

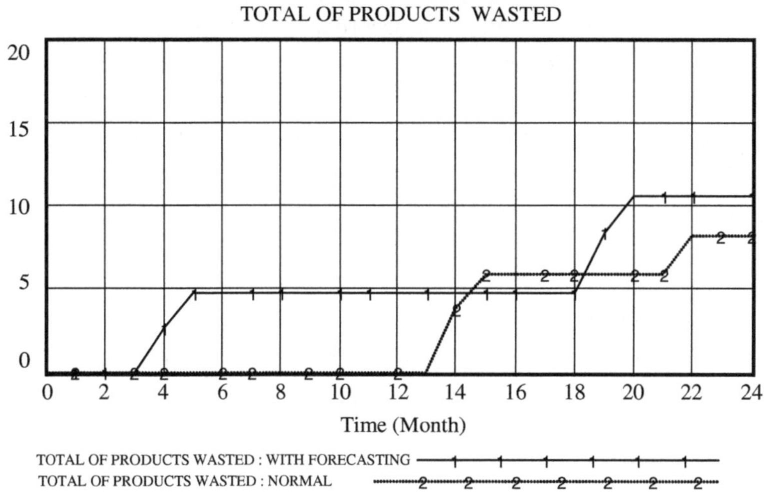

Fig. 5.21 Total of *Products wasted*

(14) ONE = INVENTORY CONTROL

(15) PERISHABLE PRODUCTS = IF THEN ELSE (COUNTER > MONTHS WITHOUT WAREHOUSE ROTATION, WAREHOUSE * 0.25, 0)

(16) PRODUCTION = IF THEN ELSE (EXPONENTIAL SMOOTH FORECASTING < WAREHOUSE, MAX (5, EXPONENTIAL SMOOTH FORECASTING), IF THEN ELSE (WAREHOUSE < 10, MAX(5, EXPONENTIAL SMOOTH FORECASTING), 5))

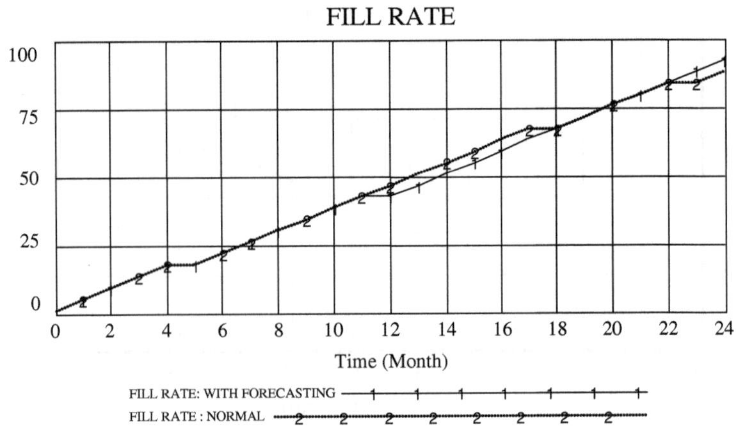

Fig. 5.22 Simulations for *Fill rates* (*Replenishment order* equal to 7 tons per period)

The modeled manufacturing order considers the inventory level in the warehouse, and manufacturing is done in accordance with the sales forecasting and with the minimum manufacturing level, which is 5 tons. In this way, the inventory does not increase in excess, while backlogged orders or stockouts lower.

(17) SALES = IF THEN ELSE (WAREHOUSE >= CUSTOMER DEMAND, CUSTOMER DEMAND, 0)

(18) SAVEPER = TIME STEP

(19) TIME STEP = 1

The time step for the simulation.

(20) TONS OF PRODUCTS WITHOUT ROTATION = 10

(21) TOTAL PRODUCTS WASTED = PERISHABLE PRODUCTS
Initial Value: 0
This variable indicates the total amount of products wasted during the simulated period.

(22) WAREHOUSE = FLOW OF PRODUCTS TO WAREHOUSE−PERISHABLE PRODUCTS−SALES
Initial Value: 20

In this case, the fill rate reached is 92.66% (Fig. 5.20):

The increase in *Products wasted* is not in excess if compared with those products wasted when using the original replenishments order (see Fig. 5.21). However, the fill rate has increased.

What will happen if the replenishment order is fixed at the maximum tons per period that the company is able to manufacture (that is, between 5 and 7 units)? Analyze the graphs below (Figs. 5.22 and 5.23).

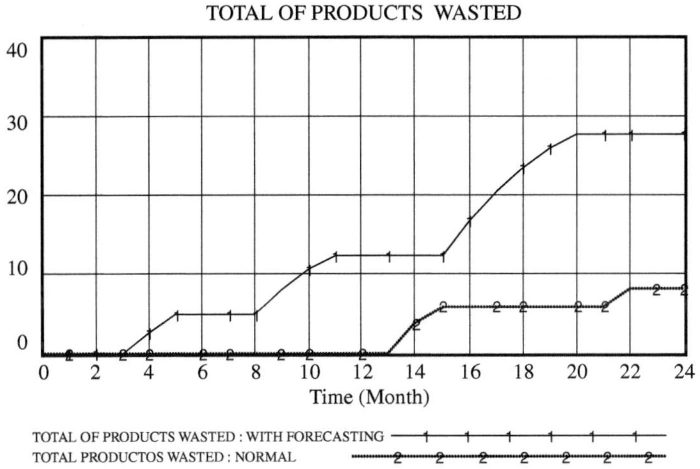

Fig. 5.23 Total of *Products Wasted* (*Replenishment Order* equal to 7 tons per period)

The fill rate reached now with the new manufacturing order is 88.5% and there are more products wasted.

5.4 Conclusions

This chapter has introduced the reader to design supply chain simulation models which, although not complex, include a series of basic variables to simulate demand management along any supply chain. The first two proposed problems model warehouse management simply with and without differed demand by using most of the variables proposed in these two problems to then move on to the third problem. Problem 3 represents a company which manufactures perishable products, whose modeling proves more complex, but prepares the reader to construct a multilevel supply chain, which Chap. 6 of this book covers.

References

M.S. Amir, Industrial viewpoint. Can systems dynamics be effective in modelling dynamic business systems? Bus. Process. Manage. J. **11**, 612–615 (2005)

J. Forrester, *Industrial Dynamics* (MIT Press, Cambridge, 1961)

J.D. Sterman, *Business Dynamics: Systems Thinking and Modeling for a Complex World* (McGraw-Hill Higher Education, New York, 2000)

Chapter 6
Modeling a Traditional Supply Chain by Using Causal Loop Diagrams

6.1 Introduction

The supply chain includes business processes, people, organization, technology and physical infrastructure, which enable the transformation of raw materials into products and intermediate and finished services that are offered and distributed to the consumer to meet demand.

The supply chain of a company includes functional areas, of both external and internal kinds, which range from the suppliers of raw materials to end consumers. All these areas will make up a level or link of the chain which will exchange materials and information with the adjacent levels.

A traditional supply chain is made up of several members. Each member receives information about the demand from the level situated immediately downstream and transmits it as orders to the level situated immediately upstream.

This chapter proposes systems dynamics-based simulation with a traditional supply chain based not only on the Inventory Order-Based Production Control System (IOBPCS) developed by Towill (1982), but also on the model proposed by Forrester (1961), and depending on its evolution in accordance with other authors (John et al.1994), this being the APIOBPCS (Automatic Pipeline, Inventory and Order-Based Production Control System) model. Chapter 4 of this book addresses both the IOBPCS and APIOBPCS models. This chapter considers a traditional supply chain with four levels: final customer, retailer, wholesaler and manufacturer. For this purpose, it considers causal loop diagrams of each level, and describes materials and information flows up- and downstream, respectively, along this traditional supply chain.

F. Campuzano and J. Mula, *Supply Chain Simulation*,
DOI: 10.1007/978-0-85729-719-8_6, © Springer-Verlag London Limited 2011

6.2 Patterns Used to Propose the Construction of a Traditional Supply Chain Model

It is necessary to follow two patterns to model and simulate the traditional supply chain. The first involves creating the causal loop diagram, while the second implies creating a flow diagram, essential for performing simulation.

The basic model created is done with a traditional supply chain whose structure is linear and is formed by the final customer, retailer, wholesaler and manufacturer levels.

The steps followed to create the causal loop diagram for the specific case of a traditional supply chain are based on the proposals of Sterman (2000) and Martín (2006), as follows:

1. Firstly, place the elements with an influence on the problem to be studied. In this case, the elements considered to create the causal loop diagram of the selected supply chain, and based on the APIOBPCS model, are the following:

   (a) *Final customer demand* and demand from one level toward the level situated immediately upstream.

   (b) *Firm orders* (for retailer, wholesaler and manufacturer). Firm orders will consist in the demand sent by the level immediately downstream of that being considered, and in the backlogs of the concerned chain echelon. In other words, if subindex i corresponds to the chain level we are considering, D_{i-1} to the demand of the level immediately downstream, and Pp_i to the backlogs of the relevant level, then firm orders will be:

   $$Firm\,orders_i = D_{i-1} + Pp_i \tag{6.1}$$

   (c) *Backlogged orders* (for retailer, wholesaler and manufacturer).

   (d) The *On-hand inventory* (for retailer, wholesaler and manufacturer). This is the inventory that can be in the warehouse, and the on-hand amount of it can never be negative. This amount is important because it determines whether a certain customer's demand can be satisfied directly from the warehouse.

   (e) *Demand forecasting* (for retailer, wholesaler and manufacturer). Forecasting has been made using exponential smoothing forecasting.

   (f) *Inventory position* (for retailer, wholesaler and manufacturer). The inventory position is defined by the following relation (Silver et al. 1998):

   $$\begin{aligned} Inventory\,position = {} & Inventory\,on\,hand \\ & + orders\,placed\,but\,not\,yet\,received \\ & (or\,on\text{-}order\,products) - backlogged\,orders \end{aligned} \tag{6.2}$$

   (g) *Replenishment orders* (for both retailer and wholesaler).

   (h) *Manufacturing orders* (manufacturer level). Both replenishment and manufacturing orders to be made according to the inventory policy chosen to manage demand. Regardless of the policy followed, the variables *Demand*

Forecasting, *Inventory Position* and *Supply or Manufacturing lead times* are taken into account to trigger these orders.
The ordering policy we have chosen for our analysis is a generalized order-up-to policy (Silver et al. 1998). In any order-up-to policy, ordering decisions are as follows:

$$O_t = S_t - Inventory\,position \tag{6.3}$$

The order quantity is equal to S_t, reduced for the inventory position or (6.2). Where O_t is the ordering decision made at the end of period t, S_t is the order-up-to level used in period t and the inventory position equals the net stock plus on order (orders placed but not yet received), and the net stock equals inventory on hand minus backlog. The order-up-to level is updated every period according to:

$$S_t = \hat{D}_t^L + k\hat{\sigma}_t^L \tag{6.4}$$

where S_t is equal to the estimate mean of demand $\overset{\wedge L}{D_t}$ over L periods $\left(\overset{\wedge L}{D_t} = \overset{\wedge}{D_t} \cdot L\right)$, increased for the prescribed fill rate with buffer stocks, σ_t^L is an estimation of the standard deviation over L periods, and k is the fill rate factor (safety factor) which depends on demand distribution (here it is assumed to be distributed normally).
In this policy, level S is updated in each period because demand forecasting is also updated. The discrepancy between the inventory position and level S will immediately become a replenishment order for the purpose of always maintaining the inventory position at level S.

(i) *Lead time* (for both wholesaler and manufacturer).
(j) *On-order products* (for retailer, wholesaler and manufacturer). It is made up of the inventory that has been served and will not be on hand until the stipulated lead time has elapsed, and of the inventory that is on hand in the warehouse after completing the manufacturing process.
(k) *Manufacturing capacity* (manufacturer level). To be expressed as the number of units that can be made in a period.
(l) *Manufacturing* (manufacturer level).
(m) *Manufacturing lead time* (manufacturer level).
(n) *Fill rates* (for retailer, wholesaler and manufacturer). Fill rates are defined as the quotient between the number of units shipped to customers on time and the total number of units demanded by them.
(o) *Inventory costs* (holding and order costs) (for retailer, wholesaler and manufacturer) and stockout costs (generated when an order is not delivered on time).

Logically, these elements vary according to the type of supply chain being modeled.

2. Next, the relationships or influences among them are defined, or more specifically, are drawn.
3. All the loops created when modeling the system must be particularly taken into account, especially the positive ones as they destabilize the system.
4. After following the aforementioned steps, it is recommendable to eliminate the elements not believed to have an influence on the system.

6.3 A Traditional Supply Chain Modeling Proposal: Causal Loop Diagram

Before explaining the steps to follow to create the causal loop diagram, it is useful to provide a description of traditional supply chain operation. Special emphasis is placed on distinguishing the information flows among the different chain levels and when materials flows take place (only finished products in this case) between each link. This provides a better understanding of the causal loop diagram.

6.3.1 Physical Description of the Traditional Supply Chain

The traditional supply chain to be modeled is made up, in this particular case, of four members: the final customer, retailer, wholesaler and manufacturer. Each one can receive orders from the member located immediately on the previous chain link, and can supply finished products at the same time (except the final customer) (see Fig. 6.1). It is worth explaining that the factory can be the first link of a different supply chain to the one under study herein. This factory's aim may be to deliver this chain's raw material, materials which have undergone several transformation processes since they were obtained until they are finally ready to be used for their assembly and subsequent sales to wholesalers.

In this case, the final customer is represented by consumers who demand the finished products that the chain supplies (pull approach). This may involve people who visit a supermarket to acquire basic articles, fungibles, or any other type of products.

The demand that these customers generate is met by the so-called retailers, represented by supermarkets, computer shops, shoe shops, etc., which acquire the articles required to perform transactions from the so-called wholesalers, represented by large wholesale warehouses supplying retailers.

The wholesaler, which could form part of the several different supply chains as it supplies several retailers, fills its warehouse with the products that the manufacturer supplies it, which is in charge of assembling and/or processing the raw materials that its suppliers deliver it to obtain finished products after the whole manufacturing and/or assembly process that end up in the final customer's hands.

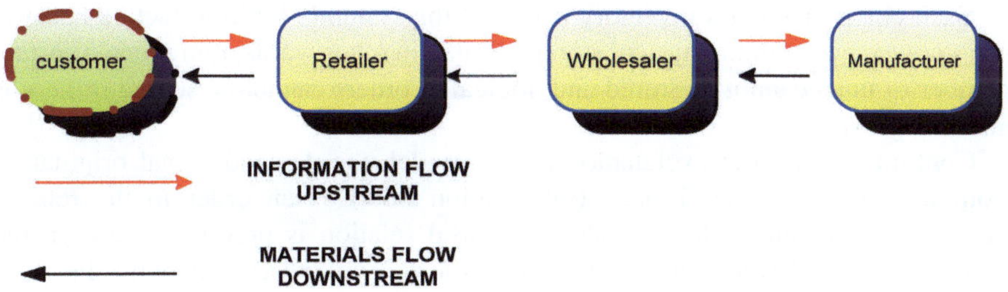

Fig. 6.1 A traditional supply chain

The information flows between all the elements forming the chain are represented by the various replenishment or manufacturing orders, while the materials flows are represented by any delivery of merchandise among these elements. No type of cooperative relationships among chain members is considered to have been established; in other words, each one is in charge of meeting the needs of the immediately previous member in the chain, and of placing orders to the member that comes immediately after.

Thus this chapter describes one part of the traditional supply chain's business process, that of demand management, delivering orders, managing the manufacturing flow and replenishment or purchases, which are used as previously mentioned to create the causal loop diagram that represents each supply chain member's demand management.

6.3.2 Causal Loop Diagram of a Traditional Supply Chain

After learning about the overall system variables that the previous section describes and the hypothetical causal relations among them, they can be graphically represented in the causal loop diagram. In this diagram, the different causal relations among the variables are represented by bows accompanied by a + or− sign that indicates the type of influence exercised by one variable on another, as Chap. 1 of this book describes.

Arrows also indicate information or materials flows, and this will be indicated later when explaining the diagram design.

The causal loop diagram begins by contemplating the lowest supply chain level: the final customer. The orders generated by this customer are delivered to the retailer, which receives them and, if possible, serves them provided its warehouse (represented by the *Inventory on-hand retailer* variable) has the amounts required by the customer available. In the causal loop diagram, the *Final customer demand* and *Firm orders retailer* variables represent these two elements.

Each chain member's inventory is one of the system's limiting factors because it cannot exceed certain physical storage limits; in other words, more than a certain number of units cannot be stored and, logically, orders cannot be served if there is not enough stock.

Continuing with the explanation of the model, the demand signal originating from the final customer is sent (information flow) as an order to the retailer (the diagram identifies how a positive causal relation is produced between the *Final customer demand* and *Backlogged orders retailer* variables with the *Firm orders retailer* variable, and how the state of this last variable increases). Firm orders are served if, as mentioned earlier, the retailer's inventory allows them to; that is, if the retailer has enough stock available to cover the amount ordered. Thus, the diagram shows that the *Firm orders retailer* variable produces a negative causal relation with the *Inventory on-hand retailer* variable as it is assumed that a possible delivery of products lowers the inventory level. If, indeed, the warehouse has the amount of finished products to meet the final customer's demand, they are delivered to the final customer, which brings about a materials flow to the final customer. If, on the other hand, the retailer warehouse does not have enough stock, the order remains to be served as soon as the warehouse has the indicated level to do so. Thus, the backlogged orders diagram forms part of the *Backlogged orders retailer* variable, which has a negative causal relation with the *Inventory on-hand retailer* variable as the lowered inventory level may increase backlogged orders since there is not enough stock to meet demand at a given time.

The orders received from the final customer are examined by the retailer and, as mentioned earlier, are delivered if there is enough inventory. These orders, in turn, form part of the data with which the forecastings of future orders are made (this demand forecasting is represented in the causal loop diagram by the *Demand forecasting retailer* variable) and will be the information that the retailer bears in mind when it places its orders. The *Final customer demand* variable has a positive causal relation with the *Demand forecasting retailer* variable as it may imply that its state increases or decreases. Whenever required, an increase in demand forecastings makes the replenishment order size greater (thus, the causal relation between these two variables is positive).

To avoid stockouts, the warehouse can have a minimum safety stock available (depending on the inventories policy followed) whose size depends on the demand forecastings and on the wholesaler lead time; that is, the time it takes the retailer to receive an order since it launches the replenishment order and this is served in the retailer warehouse.

In the causal loop diagram, the *Inventory position retailer* variable controls the inventory level, which is the function of the *Inventory on-hand retailer*, *On-order products retailer* and *Backlogged orders retailer* variables. With the first two variables, the causal relation is positive as any increment in them increases the inventory position. Increased backlogged orders lowers the inventory position, thus the causal relation with the *Inventory position retailer* variable is negative.

When considering systems dynamics, the minimum inventory level required to avoid stockouts is that which Chap. 4 has defined as the position desired and which

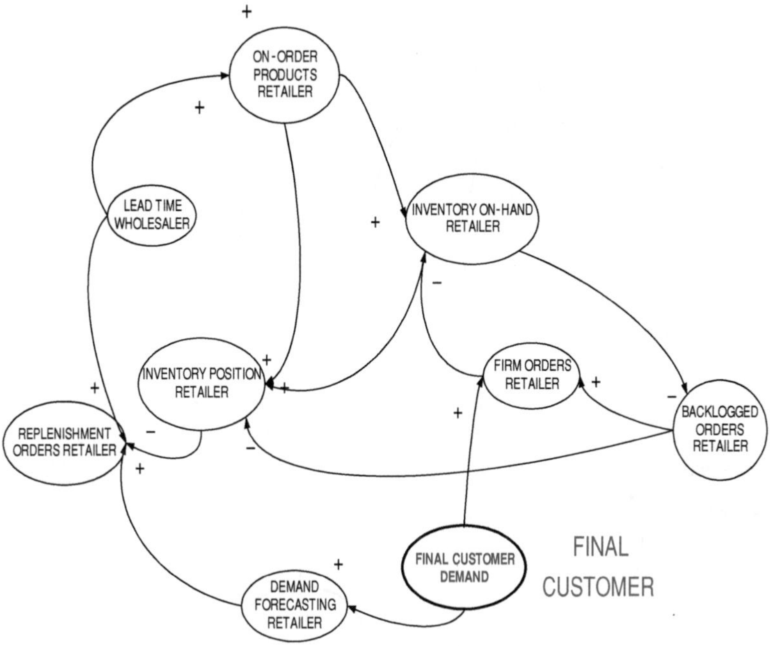

Fig. 6.2 Causal loop diagram illustrating the retailer level. A traditional supply chain

the system compares at all times with the current inventory position, which represents the real system state. The action needed to lead the system to the desired state, at least, is determined by the replenishment order.

When the retailer has all the information needed available to launch an order, that is, demand forecastings, its current inventory position and the lead time wholesaler (in the causal loop diagram, the *Replenishment orders retailer* variable has the *Inventory position retailer*, *Demand forecasting retailer* and *Lead time wholesaler* variables as inputs), it generates the order that reaches the wholesaler (information flow) after a period of time, and depends on how both parties use ICT. Delays in information among levels, and the lack of clear and reliable information, may bring about the bullwhip effect, which Chap. 3 of this book addresses. Figure 6.2 reflects the retailer level in detail.

The *Lead time wholesaler* and *Demand forecasting retailer* variables have a positive causal relation with the *Replenishment orders retailer* variable as any increase in them augments the last variable. The same does not occur with the *Inventory position retailer* variable as this conditions the number and size of the replenishment orders generated.

At the wholesaler level, the replenishment order received from the retailer brings about the same process as that described earlier. The orders generated by the retailer are sent to the wholesaler, which receives them and, if possible, delivers them in accordance with the amounts required by the retailer being available in its

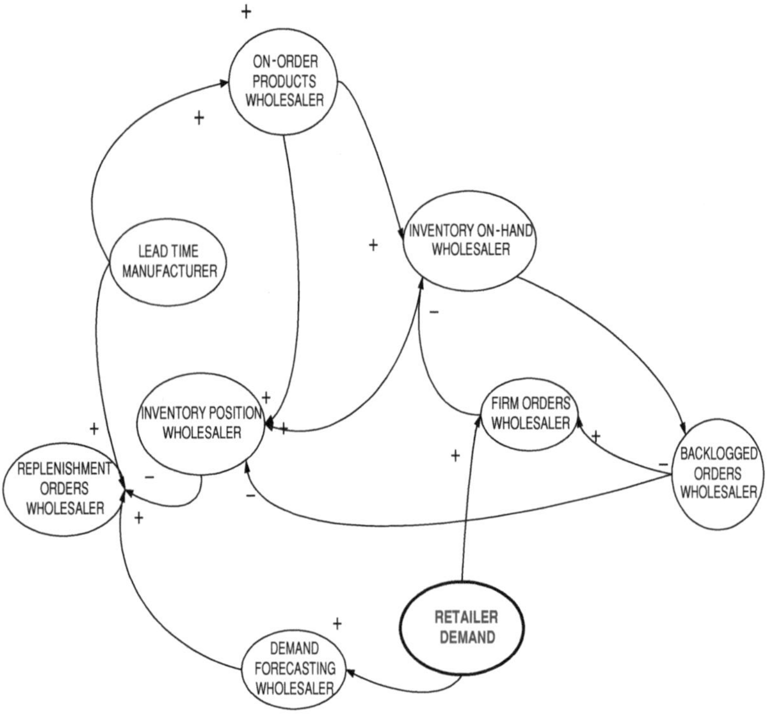

Fig. 6.3 Causal loop diagram of the wholesaler level. Traditional supply chain

warehouse. In the diagram, these two elements are represented by the *Retailer demand* variable, which receives the information from the retailer level thanks to the *Replenishment orders retailer* and *Firm orders wholesaler* variables.

As mentioned earlier, the demand signal originating from the retailer is sent (information flow) as an order to the wholesaler. The diagram of Fig. 6.3 shows how a positive causal relation is produced between the *Retailer demand* and *Backlogged orders wholesaler* variables with the *Firm orders wholesaler* variable as the state of the last variable has increased. Firm orders are served if the wholesaler inventory position permits this; that is, if there is enough stock in its warehouse to meet the amount of the demand. Therefore, the diagram depicts how the *Firm orders wholesaler* variable produces a negative causal relation with the *Inventory position wholesaler* variable, which implies a possible delivery of products that lowers the inventory level.

If indeed the wholesaler warehouse has the amount of finished products available to meet the retailer's demand, these articles are delivered to the retailer, which brings about a materials flow toward the retailer. If, on the other hand, the wholesaler warehouse does not have enough stock, the order remains to be served as soon as the warehouse has the indicated level to do so. Thus, the backlogged orders in the diagram form part of the *Backlogged orders wholesaler* variable, which has a negative causal relation with the *Inventory position wholesaler* variable as a drop in the inventory level may increase backlogged orders because there is insufficient stock to meet demand at a given time.

The orders delivered by the retailer are not only met if there is sufficient inventory, but also form part of the information with which the forecastings of future orders are made. These demand forecastings are represented in the diagram by the *Demand forecasting wholesaler* variable, which is information that the wholesale bears in mind when placing its orders. The *Demand retailer* variable has a positive causal relation with the *Demand forecasting wholesaler* variable as it increases its state. Whenever necessary, an increase in the demand forecastings makes the replenishment order size larger (therefore the causal relation between the two variables is positive). Thus, as with the retailer, when the wholesaler has all the required information to launch an order, that is demand forecastings, its current inventory position and the lead time wholesaler (in the causal loop diagram, the *Replenishment orders wholesaler* variable has the *Inventory position wholesaler*, *Demand forecasting wholesaler* and *Lead time manufacturer* variables as input) generate the order that will reach the manufacturer (information flow) in a time that depends on the two parties' use of ICT.

The increase (both positive and negative) in demand variability can be noteworthy at this level given factors like variable lead times or inaccurate forecastings which alter the original demand signal.

The figure below depicts the wholesaler level in detail (Fig. 6.3).

The manufacturer receives orders from the wholesaler and, as explained earlier for the other levels, these orders are met provided its warehouse has enough units that the wholesaler requires. In the diagram, these two elements are represented by the *Demand wholesaler* variable, which receives information from the retailer level thanks to the *Replenishment orders wholesaler* and *Firm orders manufacturer* variables. As previously mentioned, the demand signal originating from the wholesaler is sent (information flow) as an order to the manufacturer. The diagram identifies how a positive causal relation is produced between the *Demand wholesaler* and *Backlogged orders manufacturer* variables with the *Firm orders manufacturer* variable, because the state of this last variable has increased. Firm orders are served if the manufacturer's inventory position permits this, that is, if it has enough stock in its warehouse to meet the amount ordered. Therefore, the diagram shows how the *Firm orders manufacturer* variable produces a negative causal relation with the *Inventory position manufacturer* variable which assumes a possible delivery of products that lowers the inventory level.

If the manufacturer's warehouse has the amount of finished products available to meet the wholesaler demand, these articles can be delivered to the wholesaler, which leads to a materials flow toward the wholesaler. If, conversely, the manufacturer warehouse does not have sufficient stock, the order will remain pending to be served until the warehouse has a sufficient level to serve it. Thus the diagram depicts the backlogged orders to form part of the *Backlogged orders manufacturer* variable, which shows a negative causal relation with the *Inventory position manufacturer* variable as a lowered inventory level can increase the backlogged orders as there is not enough stock to cover demand at a particular time.

The orders received from the wholesaler are examined and are met if there is enough stock, and also form part of the data with which the forecastings of future orders are made. This demand forecasting is represented in the diagram by the *Demand forecasting manufacturer* variable and is information that the manufacturer bears in mind to place its orders. The *Demand wholesaler* variable has a positive causal relation with the *Demand forecasting manufacturer* variable as the state of the latter variable increases. Whenever necessary, an increase in demand forecastings makes the size of the replenishment order larger, therefore the causal relation between these two variables is positive.

When the manufacturer has all the information needed to launch an order to the factory, that is, demand forecastings, its current inventory position and the manufacturing lead time (in the causal loop diagram, the *Manufacturing orders* variable has the *Inventory position manufacturer, Demand forecasting manufacturer* and *Manufacturing lead time* variables as input) generate an order that is met according to the factory's capacity; that is, capacity restrictions such as the number of operators available, the number of hours per week according to the collective agreement, the machinery's operation capacity (the number of parts per unit of time manufacturing). They are limiting factors for the system, and are also one of the system's key factors.

The manufacturing capacity will be, on the one hand, a limiting factor for the system since not all the units required to cover demand can be produced on each occasion, which leads to unmet orders along the chain. On the other hand, it is also a key factor as any amendment made to this system may alter the fill rates along the chain.

The figure below presents details of the manufacturer level (Fig. 6.4).

The cause-and-effect relations among the variables constituting the causal model of a traditional supply chain have been defined. All that remains to remark upon is that each supplier's lead time to its corresponding customer conditions the materials flows among the levels. At the retailer level, the *On-order products retailer* variable is conditioned by the lead time wholesaler which, in turn, conditions the delivery of products to its warehouses by the manufacturer warehouse, which has to wait if it does not have sufficient stock until the products have been manufactured to be delivered. This also influences the delivery of products to both the retailer and the final customer.

Lead times are also a key factor for the system as any amendment made to them implies alterations to inventories, replenishment orders or to manufacturing, costs, etc.

6.4 Other Supply Chain Management Areas

Nowadays, companies are confronted with many challenges. Growing competitiveness and globalization require ever-efficient responses and solutions (processes and strategies) of companies that enable them to interact in a continuously

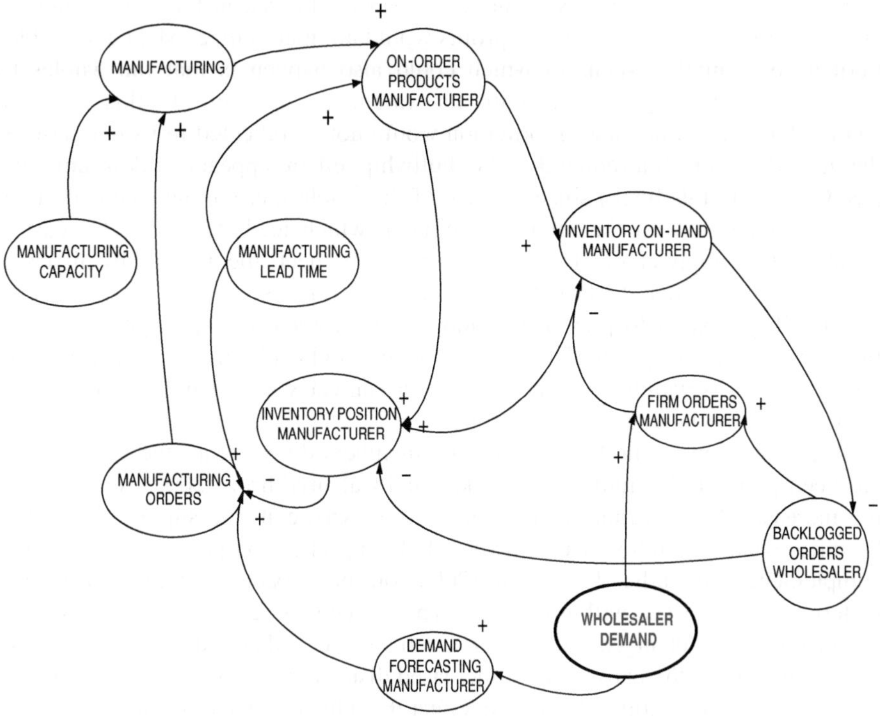

Fig. 6.4 Causal loop diagram of the manufacturer level. A traditional supply chain

changing world, where costumers increasingly hold bargaining power and finally determine the success or failure of the whole business mechanism which underlies the manufacturing of a given product. This scenario becomes even more complex if we consider that the production and marketing work do not finish at the time a sale is made; only after the client has accepted is it wholly satisfied with the product and pays; all this forward manufacturing flow has fulfilled its mission, regardless of the closed-loop or reverse supply chain model.

Adequate supply chain management is crucial for companies to remain competitive. Currently, competition appears not only among companies, but also among supply chains; new management tools (based on ICT) are fostering the integration of companies into supply chains, and the emergence of organizations capable of responding more efficiently. Nevertheless, some underlying issues should be resolved to achieve the supply chain's efficient operation.

It is very common for wholesalers and manufacturers to establish product pricing policies. Depending on each individual case, the manufacturer can offer discounts, usually per product volume, to encourage the wholesalers to purchase a larger amount than they need. If the difference between the item's real price and the purchase price is greater than the item's holding costs, this strategy could prove initially profitable for the wholesaler, but if it has not carried out a study to know what its real demand actually is, holding costs could exceed the aforementioned price difference, thus the expected profits would not materialize (Mela et al. 1998). Furthermore, another scenario which could also happen is that the wholesaler continues purchasing products until its warehouses are full. In this case, the wholesaler's actual demand information would not be reflected in its purchases to the manufacturer. Subsequently, the bullwhip effect appears and/or amplifies (see Chap. 3 in this book) since the size of the wholesaler's replenishment orders bears no relation with the retailer's demand, which leads to wrong forecasting upstream the supply chain. The wholesaler merely meets the demand by disposing of stored products that distorts the manufacturer's forecastings, which notes a drop in sales if compared to previous periods. These forecastings prompts the manufacturer to reduce its production activity, which might subsequently result in stock disruptions in its warehouse. The wholesaler can carry out such price policies for retailers and the latter for final customers.

Price variation or modification by the members downstream the supply chain has been particularly studied by Özelkan and Çakanyildirim (2009), who analyze the increase of this variation when moving downstream the supply chain. This becomes what the authors call a reverse bullwhip effect on pricing. This analysis complements the studies by Cowan (2004) on the impact of changes in demand patterns on sale prices within different types of economies and, consequently, on the profits these changes entail. For Özelkan and Çakanyildirim (2009), price fluctuation can be the cause behind increased distortion in replenishment orders or, in other words, an amplified bullwhip effect that affects the supply chain upstream.

Campuzano et al. (2011) study the fluctuation of the retailer's prices and its impact on the proper management of a traditional multilevel supply chain (made up of a manufacturer, wholesaler, retailer and final customer). By using a staggered step demand pattern, which responds elastically to retailer price fluctuation, the influence of these fluctuations on the variability of the orders placed along the modeled supply chain is analyzed by means of a dynamic supply chain management model. In order to quantify this distortion, the coefficient of the variation associated with each series (replenishment/manufacturing orders) is calculated and obtained from the different simulations carried out with the dynamic model proposed using a seasonal price pattern that is disturbed by different variability levels. In addition, and given the collaboration environment among the different members of the chain, the use of linear regression is proposed, which is a predictive model that enables to quantify the distortion of the orders generated by using information on variability prices and the orders at the level immediately downstream.

On the other hand, discounts on the purchase price have been profusely dealt with in the current specialized literature in the operations management field by focusing (research) on models to search for the optimal strategy for retailers to follow when it comes to applying discounts to the final consumer purchase price in order to avoid inventory problems and the subsequent holding costs. The works by Arcelus and Srinivasan (1995, 1998) and Arcelus et al. (2001) or Ardalan (1998) are worthy of mention. Some use demand patterns that are sensitive to price fluctuations, but they do not achieve the simulation of their effect along a multi-level supply chain as far as the bullwhip effect is concerned.

6.5 Conclusions

Based on the APIOBPCS model, this chapter has identified the main variables used for demand management along a traditional supply chain. Once the variables have been identified, the various causal loop diagrams have been constructed for the retailer, wholesaler and manufacturer levels. The next step consists firstly in constructing the flow diagram for this supply chain by using any software that addresses systems dynamics modeling (Vensim®, for instance), and secondly in simulating the created model. The reader will be able to recognize the interdependencies among the various links of the supply chain which can, for example, lead to periods when products are scarce which favor stockouts, or even to excessive replenishment orders because of inaccurate forecastings which lead to excessive inventories and, consequently, to increased storage costs. These problems can be overcome by amending replenishment orders, lead times, or by simply adjusting forecastings. The various scenarios (what-if analysis), which enable the model to be recreated by means of systems dynamics, give an idea of this methodology potential to solve demand management problems, among others.

References

F.J. Arcelus, G. Srinivasan, Discount strategies for one-time-only sales. IIE. Trans. **27**, 618–624 (1995)

F.J. Arcelus, G. Srinivasan, Ordering policies under one-time-only discount and price sensitive demand. IIE. Trans. **30**, 1057–1064 (1998)

F.J. Arcelus, N.H. Shah, G. Srinivasan, Retailer's response to special sales: price discount vs. trade credit. Omega-Int. J. Manage. S. **29**, 417–428 (2001)

A. Ardalan, Optimal ordering policies in response to a sale. IIE. Trans. **20**, 292–294 (1998)

F. Campuzano, A. Lisec, A. Guillamón (2011) Assessing the impact of prices fluctuation on demand distortion within a multiechelon supply chain. Promet **23**, 131–140 (2011)

S. Cowan Demand shifts and imperfect competition, University of Oxford, Department of Economics Discussion Paper Series; March 2004; Number 188; ISSN 1471-0498. (2004) Available at URL: http://www.economics.ox.ac.uk/Research/wp/pdf/paper188.pdf

J. Forrester, *Industrial dynamics* (MIT. Press., Cambridge, MA, 1961)

S. John, M.M. Naim, D.R. Towill, Dynamic analysis of a WIP compensated decision support system. Int. J. Manuf. Syst. Des. **1**, 283–297 (1994)

J. Martín, *Theory and Practical Exercises of System Dynamics*, ed. by J.M. García (Barcelona, 2006)

C.F. Mela, K. Jedidi, D. Bowman, The long-term impact of promotions on consumer stockpiling behaviour. J. Marketing. Res. **35**, 250–262 (1998)

E.C. Özelkan, M. Çakanyildirim, Reverse bullwhip effect in pricing. Eur. J. Oper. Res. **192**, 302–312 (2009)

E.A. Silver, D.F. Pyke, R. Peterson, *Inventory Management and Production Planning and Scheduling* (John Wiley & Sons, New York, 1998)

J.D. Sterman, *Business Dynamics: Systems Thinking and Modeling for a Complex World* (McGraw-Hill Higher Education, New York, 2000)

D.R. Towill, Dynamic analysis of an inventory and order based production control system. Int. J. Prod. Res. **20**, 369–383 (1982)

Chapter 7
Getting into Practice: Modeling an Entire Traditional Supply Chain

7.1 Introduction

In this chapter, we construct a simulation model of a traditional supply chain. We select systems dynamics in general and the Vensim® simulation software in particular as a basis for this book to study supply chain dynamics problems. Regarding systems dynamics, we agree with Sterman (2000) and Amir (2005) about its effectiveness to model dynamic business systems, in this case, supply chains. The reason for selecting the Vensim® software is because it uses a modeling approach that combines systems dynamics and simulation concepts with discrete events to represent a supply chain's events and uncertainties in detail and to subsequently analyze its performance based on its structure and the existing causal relations among its components. Local simulation is used to model the proposed example problems.

This chapter is organized as follows: first, the characteristics and variables of the model are defined. Then, these variables are identified as level, flow and auxiliary variables to develop the Forrester's diagram of the model. Finally, the chapter describes the model in detail and formulates the corresponding equations which are explained throughout the text.

7.2 Practice Problem: Modeling an Entire Traditional Supply Chain

The reader is recommended to use Vensim® to create the flow diagram of a traditional supply chain that has two members: a retailer and a manufacturer (Fig. 7.1).

The model's characteristics (the values of the parameters employed) to be constructed for their subsequent simulation are as follows:

F. Campuzano and J. Mula, *Supply Chain Simulation*,
DOI: 10.1007/978-0-85729-719-8_7, © Springer-Verlag London Limited 2011

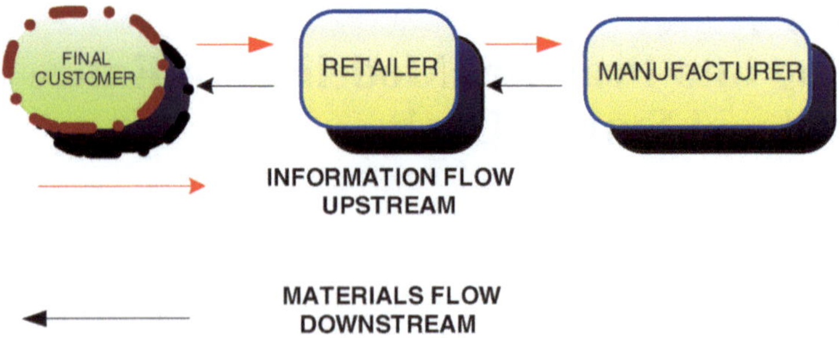

Fig. 7.1 Traditional supply chain with three levels

- The demand pattern selected corresponds to a normal distribution. Simulation takes place over 365- periods.
- The stock of the initial inventory for each level is 100 units.
- The manufacturing capacity is 160 units per period. The number of operators is 40 and each product unit requires 2 h to be manufactured. It is assumed that the factory produces during 8 h per period for 365 periods simulated.
- The manufacturing lead time to the retailer is 2 periods. This time is assumed to be constant for each order received, except for stockouts which logically vary.
- The manufacturing lead time is 2 periods.
- The fill rate of service K for each level equals 2.
- The adjust factor for forecasting is equal to 2, so $\alpha = 0.5$.

7.2.1 Observations for Constructing the Proposed Model

Chapter 4 of this book presents the variables to consider to construct the causal loop diagram proposed for the traditional supply chain, which can be modified according to the proposed supply chain structure. The variables used to construct the causal loop diagram of the traditional supply chain were:

a. *Final customer demand* and the demand of any level located toward the upstream level from it.
b. *Firm orders.*
c. *Backlogged orders.*
d. *Inventory position.*
e. *Demand forecasting.*
f. *Inventory position.*
g. *Replenishment orders.*
h. *Manufacturing orders.*
i. *Lead times.*
j. *On-order products.*
k. *Manufacturing capacity.*

l. *Manufacturing.*
m. *Manufacturing lead time.*
n. *Fill rates.*

The aforementioned variables are transferred from the causal loop diagram to the flow diagram by identifying which variables become auxiliary, level or flow variables. It is noteworthy that the necessary variables have been added to suitably create the flow diagram, and that they were not required for the causal loop diagrams (particularly the flow variables that amend level variables):

1. *Final customer demand.* These are auxiliary variables that condition the flow of the output of finished products from the warehouse (the available inventory variable). Information is supplied that is not conserved but updated during each period.
2. *Firm orders.* Firm orders are considered auxiliary variables as they supply information about the demand of those products among the various levels in the chain and the backlogged orders that are still to be served. This information is not conserved but is updated during each period. According to the warehouse position (available inventory) corresponding to the level the order arrives at, this order is either accepted and served, or becomes a backlogged order.
3. *Backlogged orders.* Backlogged orders become a level variable because the firm orders not served must be "stored". These orders are served when the warehouse has enough finished products to serve them (available inventory).
4. *Inventory on-hand* (warehouse). This is another level variable as it reflects all the finished products reaching each chain element and are available for delivery to possible customers. This is modified by the flow variables of the flow of output of finished products and by the arrival or flow of products to the warehouse.
5. *Products delivered.* This is considered a flow variable of materials which modifies the products warehouse position (level variable).
6. *Demand forecasting.* It is an auxiliary variable that supplies information about demand forecasting (orders) at the level immediately before that being considered. This information is not conserved but is updated during each period as simple exponential smoothing is used as a forecasting technique.
7. *Inventory position.* It is a vital auxiliary variable for the inventories policy which is used for demand management. It depends on the *On-order products*, *Backlogged orders*, *Inventory on-hand* and *Products delivered* variables. The inventory position is an information flow that is updated in each period.
8. *Replenishment orders.* Irrespectively of each level, they are auxiliary variables that send information about the products that each chain member requires to meet the demand forecast and any backlogged orders it may have. These replenishment orders are the demand of the level immediately above that being considered. Demand plus backlogged orders constitute firm orders.
9. *Manufacturing orders.* An exclusive auxiliary variable from the manufacturer level. It supplies information about the products that must be manufactured to meet future demands.

10. *Lead times*. Lead times are auxiliary variables that influence the arrival of products at the warehouse. They delay the materials flow, that is, the time between issuing the replenishment order until it is met depends on the stipulated or required lead time, unless other problems arise that increase this period, such as stockouts.
11. *Flow of products* to the level that has ordered a product. This is a flow variable of materials that modifies the state of the level variable related with on-order products.
12. *On-order products*. This is a level variable whose outputs are conditioned by the lead times corresponding to the chain member immediately upstream of that member which placed the order.
13. *Products delivered*. This is a flow variable of materials that is conditioned by the lead time of the chain member immediately upstream of that which placed the order. It amends the level variables *On-order products* and *Inventory on-hand* (warehouse) because, having elapsed the necessary lead time, this variable forces, if there are any, an output of products from the level variable *On-order products* and an input of products to the level variable *Inventory on-hand*.
14. *Manufacturing capacity*. An auxiliary variable that defines the manufacturing restrictions. This is, therefore, an auxiliary variable that conditions the amount of products to be made in the time unit considered.
15. *Manufacturing*. A level variable as it reflects the work in process (WIP) done in the factory. Finished products will move to the warehouse after the manufacturing lead time has elapsed.
16. *Manufacturing lead time*. An auxiliary variable that defines the time required to manufacture products.
17. *Fill rates*. The fill rate offers a measurement of the percentage of delivered orders in relation to all the orders placed by the final customer. They are level variables as they vary with time in terms of whether or not the different orders are delivered.
18. *Variance calculations*. Demand variance must be done in each period to calculate its standard deviation, which is included when calculating the replenishment and manufacturing orders. Thus, this is a level variable.

The characteristics of the constructed model are as follows:

1. The system works against the warehouse, which means it forecasts the daily products demands and, in relation to this forecasting and the inventory position, replenishment or manufacturing orders are sent. The manufacturer launches a manufacturing order whenever it needs to. The raw materials employed for manufacturing are considered to be available at all times and there are no delays in their delivery.
2. The final customer's real demand is completely random and constant in time.
3. Uniproduct. The possibility of several types of products is not contemplated for the time being.

4. The inventories policies pursued by all the chain members consists in delivering all the orders received, although there could be cases of lost demand, which means that this received order follows two stages:

   - It is received and examined. Besides, it is used to foresee future orders (information flow).
   - If it is possible, that is, if the warehouse has enough inventory of finished products, then this order is met. The possibility of serving only one part of the order when the whole order cannot be served is also modeled (a postponed amount); the possibility of losing an order owing to lack of warehouse stock has not been considered.

5. Inventories management has been performed using inventories review policies. The terminology employed from this point onward to define the inventory control policy used is as follows:

   - *Inventory on-hand*. This inventory is that found inside the warehouse, and the amount of available inventory can never be negative. This amount is relevant as it determines whether the demand of a certain customer can be met directly from the warehouse.
   - *Net inventory*. This inventory is equal to the available inventory minus any backlogged orders at a given time t. The option of receiving orders in advance is not considered as they would generate reserves on the available inventory. This amount can become negative (especially is there is a large amount of backlogged orders).
   - *Inventory position*. Inventory position is defined by Eq. 6.2. Given the Vensim® program characteristics, to update the inventory position at all times, the *Products delivered* variable in each period must be deducted from the former formulation.
   - *Safety stock*. Safety stock is protection against demand variability as the variability in the distributor's delivery times is eliminated, provided it is possible to do this in the suppliers selection phase.
     The inventory control policy used is the following (Silver et al. 1998):

   (a) Order up to level S. This policy is based on maintaining the inventory position within level S. The replenishment or manufacturing orders are sent provided the inventory position drops below level S. By way of example, S can be made to equal the demand forecast during the lead time, plus the standard deviation of demand during the lead time multiplied by the K service factor (Silver et al. 1998), see (Sect. 6.4).

6. Of all the manufacturing capacity restrictions, as number of employees, available hours, subcontracting, overtime, the machinery's capacity and maintaining the machinery are considered, it is only necessary to bear in mind the number of hours per period worked and the number of available workers,

which are constant, as well as the number of hours needed to manufacture each product (in this case, it is a uniproduct model).

7.3 Solution to the Practice Problem

Next this section presents the solution to the problem considered. Given the difficulty of the model, the reader is recommended to analyze the model solved in Vensim® and to understand its modeling before starting to construct it him/herself. Units have been omitted. The reader may use the units he/she wishes apart from those considered in the model formulation (Figs. 7.2 and 7.3).

VARIABLES IN THE MODEL

(01) AVERAGE DEMAND (RETAILER) = CUMULATIVE DEMAND 0/ COUNTER 0

(02) AVERAGE DEMAND1 = CUMULATIVE DEMAND 1/COUNTER 1

(03) "BACKLOGGED ORDERS DELIVERED (MANUFACTURER)" = IF THEN ELSE (DELIVERED PRODUCTS TO RETAILER ="FIRM ORDERS (MANUFACTURER)", "BACKLOGGED ORDERS (MANUFACTURER)", IF THEN ELSE (DELIVERED PRODUCTS TO RETAILER > RETAILER DEMAND, DELIVERED PRODUCTS TO RETAILER-RETAILER DEMAND, 0))

(04) "BACKLOGGED ORDERS DELIVERED (RETAILER)" = IF THEN ELSE (PRODUCTS DELIVERED TO FINAL CUSTOMER = "FIRM ORDERS (RETAILER)","BACKLOGGED ORDERS (RETAILER)", IF THEN ELSE (PRODUCTS DELIVERED TO FINAL CUSTOMER > FINAL CUSTOMER DEMAND, PRODUCTS DELIVERED TO FINAL CUSTOMER-FINAL CUSTOMER DEMAND, 0))

(05) "BACKLOGGED ORDERS (MANUFACTURER)" = INTEG (IF THEN ELSE ("INVENTORY ON- HAND (MANUFACTURER)" + PRODUCTS MANUFACTURED- "FIRM ORDERS (MANUFACTURER)" > = 0, 0, IF THEN ELSE (DELIVERED PRODUCTS TO RETAILER < RETAILER DEMAND, RETAILER DEMAND-DELIVERED PRODUCTS TO RETAILER, 0))-"BACKLOGGED ORDERS DELIVERED (MANUFACTURER)")

Initial value: 0

(06) "BACKLOGGED ORDERS (RETAILER)" = INTEG (IF THEN ELSE ("INVENTORY ON-HAND (RETAILER)" + PRODUCTS DELIVERED TO RETAILER-"FIRM ORDERS (RETAILER)" > = 0, 0, IF THEN ELSE (PRODUCTS DELIVERED TO FINAL CUSTOMER < FINAL CUSTOMER DEMAND,FINAL CUSTOMER DEMAND -PRODUCTS DELIVERED TO FINAL CUSTOMER, 0))-"BACKLOGGED ORDERS

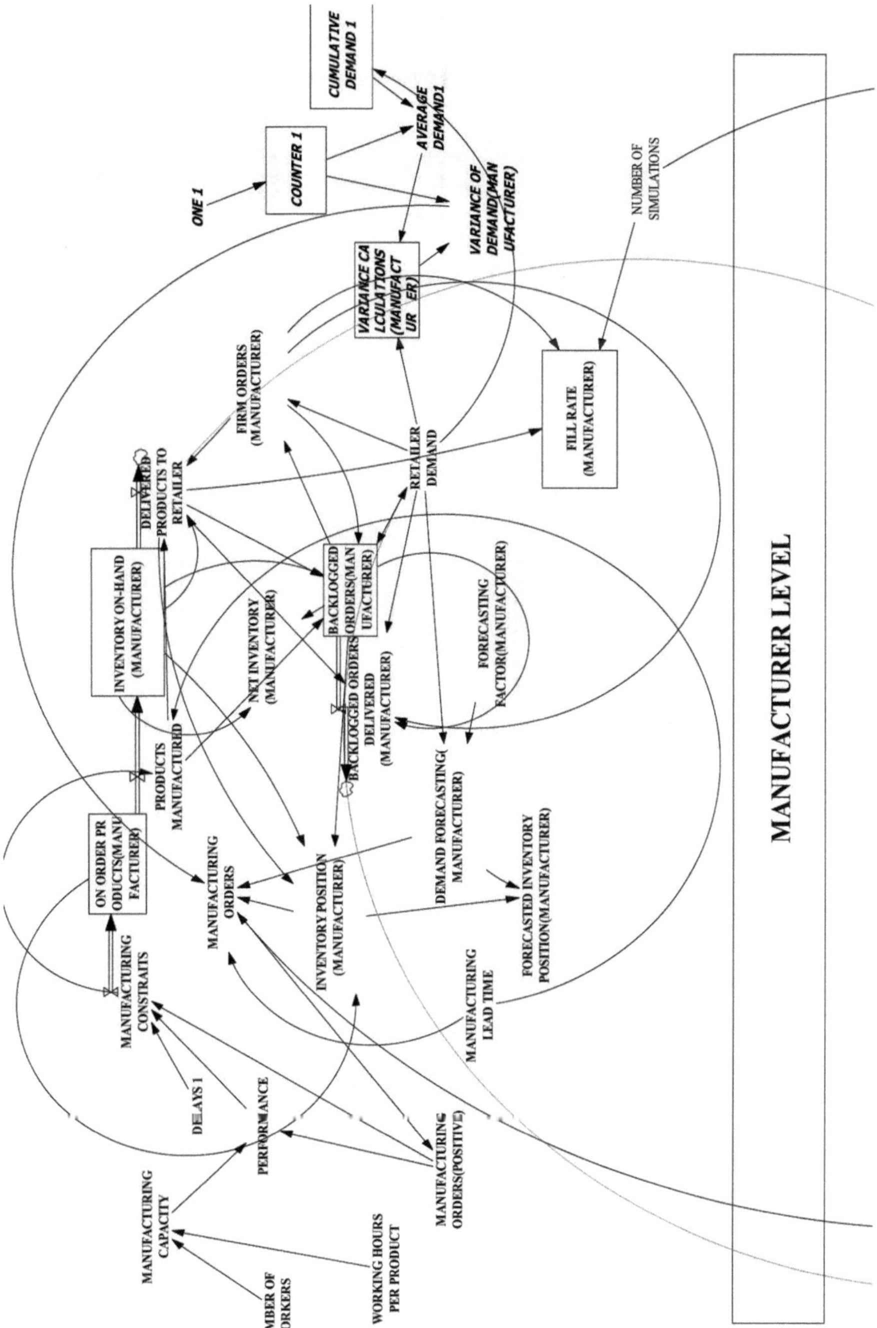

Fig. 7.2 Manufacturer level

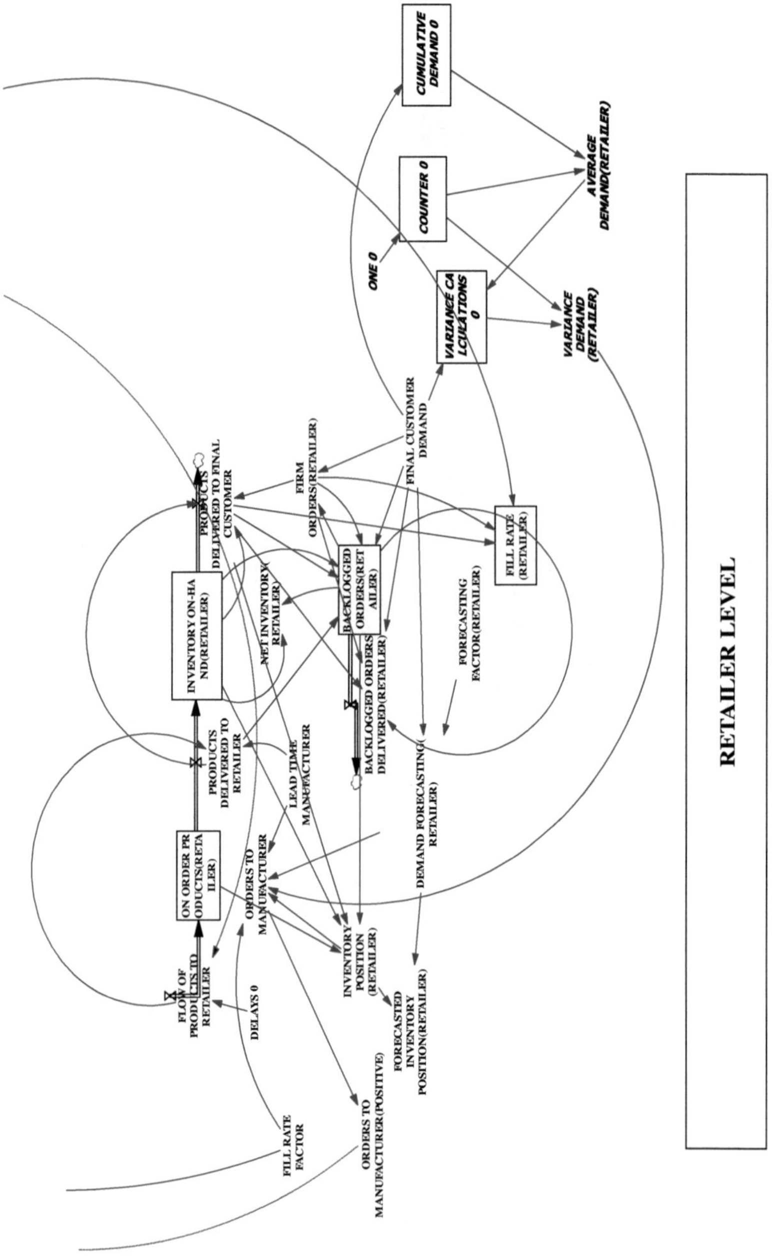

Fig. 7.3 Retailer level

DELIVERED (RETAILER)")
Initial value: 0

(07) COUNTER 0 = INTEG (ONE 0)
Initial value: ONE 0

(08) COUNTER 1 = INTEG (ONE 1)
Initial value: ONE 1

(09) CUMULATIVE DEMAND 0 = INTEG (FINAL CUSTOMER DEMAND)
Initial value: FINAL CUSTOMER DEMAND)

(10) CUMULATIVE DEMAND 1 = INTEG (RETAILER DEMAND)
Initial value: RETAILER DEMAND)

(11) DELAYS 0 = 1
(12) DELAYS 1 = 1
(13) DELIVERED PRODUCTS TO RETAILER = IF THEN ELSE (("INVENTORY-ON-HAND (MANUFACTURER)" + PRODUCTS MANUFACTURED-"FIRM ORDERS (MANUFACTURER)") > = 0,"FIRM ORDERS (MANUFACTURER)","INVENTORY ON-HAND (MANUFACTURER)")
(14) "DEMAND FORECASTING (MANUFACTURER)" = SMOOTH (RETAILERDEMAND, "FORECASTING FACTOR (MANUFACTURER)")
(15) "DEMAND FORECASTING (RETAILER)" = SMOOTH (FINAL CUSTOMER DEMAND, "FORECASTING FACTOR (RETAILER)")
(16) "FILL RATE (MANUFACTURER)" = INTEG ((XIDZ (DELIVERED PRODUCTS TO RETAILER,"FIRM ORDERS (MANUFACTURER)", 1)/ NUMBER OF SIMULATIONS)*100)

Initial value: (XIDZ (DELIVERED PRODUCTS TO RETAILER, "FIRM ORDERS (MANUFACTURER)",1)/NUMBER OF SIMULATIONS)*100)

(17) "FILL RATE (RETAILER)" = INTEG ((XIDZ(PRODUCTS DELIVERED TO FINAL CUSTOMER,"FIRM ORDERS (RETAILER)", 1)/NUMBER OF SIMULATIONS)*100)

Initial value: (XIDZ (PRODUCTS DELIVERED TO FINAL CUSTOMER,"FIRM ORDERS (RETAILER)",1)/NUMBER OF SIMULATIONS)*100)

(18) FILL RATE FACTOR = 2
(19) FINAL CUSTOMER DEMAND = PULSE (1, 2)*0 + PULSE (2,365)*RANDOM NORMAL (5, 20, 15, 8, 99)
(20) FINAL TIME = 365
(21) "FIRM ORDERS (MANUFACTURER)" = RETAILER DEMAND + "BACKLOGGED ORDERS (MANUFACTURER)"
(22) "FIRM ORDERS (RETAILER)" = FINAL CUSTOMER DEMAND + "BACKLOGGED ORDERS (RETAILER)"
(23) FLOW OF PRODUCTS TO RETAILER = DELIVERED PRODUCTS TO RETAILER/DELAYS 0

(24) "FORECASTED INVENTORY POSITION (MANUFACTURER)" = "INVENTORY POSITION (MANUFACTURER)"-"DEMAND FORECASTING (MANUFACTURER)"
(25) "FORECASTED INVENTORY POSITION (RETAILER)" = "INVENTORY POSITION (RETAILER)"-"DEMAND FORECASTING (RETAILER)"
(26) "FORECASTING FACTOR (MANUFACTURER)" = 2
(27) "FORECASTING FACTOR (RETAILER)" = 2
(28) INITIAL TIME = 1
(29) "INVENTORY ON-HAND (MANUFACTURER)" = INTEG (PRODUCTS MANUFACTURED-DELIVERED PRODUCTS TO RETAILER)
Initial value: 100
(30) "INVENTORY ON-HAND (RETAILER)" = INTEG (PRODUCTS DELIVERED TO RETAILER-PRODUCTS DELIVERED TO FINAL CUSTOMER)
Initial value: 100
(31) "INVENTORY POSITION (RETAILER)" = "INVENTORY ON-HAND (RETAILER)" + "ON ORDER PRODUCTS (RETAILER)"-"BACKLOGGED ORDERS (RETAILER)" -PRODUCTS DELIVERED TO FINAL CUSTOMER
(32) "INVENTORY POSITION (MANUFACTURER)" = "INVENTORY ON-HAND (MANUFACTURER)"-DELIVERED PRODUCTS TO RETAILER + "ON ORDER PRODUCTS(MANUFACTURER)" -"BACKLOGGED ORDERS (MANUFACTURER)"
(33) LEAD TIME MANUFACTURER = 2
(34) MANUFACTURING CAPACITY = 8*NUMBER OF WORKERS/WORKING HOURS PER PRODUCT
(35) MANUFACTURING CONSTRAITS = IF THEN ELSE (PERFORMANCE < 1,"MANUFACTURING ORDERS (POSITIVE)"*PERFORMANCE/DELAYS 1,"MANUFACTURING ORDERS (POSITIVE)"/DELAYS 1)
(36) MANUFACTURING LEAD TIME = 2
(37) MANUFACTURING ORDERS = (MANUFACTURING LEAD TIME* "DEMAND FORECASTING (MANUFACTURER)" + (FILL RATE FACTOR* (SQRT ("VARIANCE OF DEMAND (MANUFACTURER)" *MANUFACTURING LEAD TIME))))-"INVENTORY POSITION (MANUFACTURER)"
(38) "MANUFACTURING ORDERS (POSITIVE)" = MAX (MANUFACTURING ORDERS, 0)
(39) "NET INVENTORY (MANUFACTURER)" = "INVENTORY ON-HAND (MANUFACTURER)"-"BACKLOGGED ORDERS (MANUFACTURER)"
(40) "NET INVENTORY (RETAILER)" = "INVENTORY ON-HAND (RETAILER)"-"BACKLOGGED ORDERS (RETAILER)"
(41) NUMBER OF SIMULATIONS = 365

(42) NUMBER OF WORKERS = 40

(43) "ON ORDER PRODUCTS (MANUFACTURER)" = INTEG (MANUFACTURING CONSTRAITS-PRODUCTS MANUFACTURED)
Initial value: 0

(44) "ON ORDER PRODUCTS (RETAILER)" = INTEG (FLOW OF PRODUCTS TO RETAILER-PRODUCTS DELIVERED TO RETAILER)
Initial value: 0

(45) ONE 0 = 1

(46) ONE 1 = 1

(47) ORDERS TO MANUFACTURER = (LEAD TIME MANUFACTURER*"DEMAND FORECASTING (RETAILER)" + (FILL RATE FACTOR* (SQRT ("VARIANCE DEMAND (RETAILER)"*LEAD TIME MANUFACTURER))))-"INVENTORY POSITION (RETAILER)"

(48) "ORDERS TO MANUFACTURER (POSITIVE)" = MAX (ORDERS TO MANUFACTURER, 0)

(49) PERFORMANCE = XIDZ (MANUFACTURING CAPACITY,"MANUFACTURING ORDERS (POSITIVE)", 0)

(50) PRODUCTS DELIVERED TO FINAL CUSTOMER = IF THEN ELSE (("INVENTORY ON HAND (RETAILER)" + PRODUCTS DELIVERED TO RETAILER -"FIRM ORDERS (RETAILER)") > = 0,"FIRM ORDERS (RETAILER)","INVENTORY ON-HAND (RETAILER)")

(51) PRODUCTS DELIVERED TO RETAILER = DELAY FIXED (FLOW OF PRODUCTS TO RETAILER, LEAD TIME MANUFACTURER, 0)

(52) PRODUCTS MANUFACTURED = DELAY FIXED (MANUFACTURING CONSTRAITS, MANUFACTURING LEAD TIME, 0)

(53) RETAILER DEMAND = "ORDERS TO MANUFACTURER (POSITIVE)"

(54) SAVEPER = TIME STEP

(55) TIME STEP = 1

(56) VARIANCE CALCULATIONS 0 = INTEG (((FINAL CUSTOMER DEMAND-"AVERAGE DEMAND (RETAILER)")^2))
Initial value: ((FINAL CUSTOMER DEMAND-"AVERAGE DEMAND (RETAILER)")^2))

(57) "VARIANCE CALCULATIONS (MANUFACTURER)" = INTEG (((RETAILER DEMAND-AVERAGE DEMAND1)^2))
Initial value: ((RETAILER DEMAND-AVERAGE DEMAND1)^2))

(58) "VARIANCE DEMAND (RETAILER)" = (VARIANCE CALCULATIONS 0/COUNTER 0)

(59) "VARIANCE OF DEMAND (MANUFACTURER)" = ("VARIANCE CALCULATIONS (MANUFACTURER)"/COUNTER 1)

(60) WORKING HOURS PER PRODUCT = 2

Next this chapter describes how the flow diagram of the supply chain proposed for the former problem is constructed.

At level 1 (final customer), customers send their orders to the retailer in accordance with their requirements. The frequency of these orders in the proposed model is daily and random. To generate this demand, it is necessary to provide the simulation software with an instruction that implies the output of a value throughout the simulation period which represents the final customer's demand.

For this level, the decision was made to create a function that provides a series of random values. This is the function decided upon:*Random Normal (m, n, p, s)* It generates a series of random values whose characteristics are defined by values *m*, *n*, *p* and *s*. Thus, demand takes a minimum *m* value, a maximum *n* value, and all the values obtained are centered on a mean *p* value with a standard deviation of *x*, where *s* is the calculation parameter of the random numbers.

At level 2 (retailer), the retailer firstly receives orders from the final customer and serves this customer depending on its inventory position. If these orders are not met by the required date, either they are served when the retailer has sufficient stock available or the order is lost. The constructed model has considered that the orders not delivered on time (backlogged orders) can be delivered later, so they can be included in the daily firm orders (if there are any).

In this way, possible orders are represented by the variable *Firm orders* (retailer), which has been formulated as shown below:

FINAL CUSTOMER DEMAND + "BACKLOGGED ORDERS (RETAILER)"

If the retailer warehouse has enough products to meet the firm order, a materials flow to the final customer takes place, defined by the flow variable *Products delivered to final customer,* which is defined in the model as follows:

IF THEN ELSE ((“INVENTORY ON-HAND (RETAILER)” + PRODUCTS DELIVERED TO RETAILER-“FIRM ORDERS (RETAILER)”) > = 0,“FIRM ORDERS (RETAILER)”,“INVENTORY ON-HAND (RETAILER)”)

This variable is subject to the conditional function, *IF THEN ELSE*, which operates as follows:

The retailer warehouse is represented by the *Inventory on-hand (retailer)* variable at which the flow of products arrives (materials flow) through the *Products delivered to retailer* variable**.** If the amount of products in the inventory at the time the order is made suffices to meet this order, then the order is delivered. If, however, the warehouse cannot meet the backlogged orders and the on-hand demand, the system hands over all the stock there is in the warehouse; should this action not be feasible, then the firm order is not delivered, and the final customer's demand is transferred to increase the possible backlogged orders, defined by the level variable *Backlogged orders (retailer),* which is defined as shown below:

IF THEN ELSE (“INVENTORY ON-HAND (RETAILER)” + PRODUCTS DELIVERED TO RETAILER-“FIRM ORDERS (RETAILER)” > = 0,0,
IF THEN ELSE (PRODUCTS DELIVERED TO FINAL CUSTOMER < FINAL CUSTOMER DEMAND,FINAL CUSTOMER DEMAND-PRODUCTS DELIVERED TO FINAL CUSTOMER, 0)) -“BACKLOGGED ORDERS DELIVERED (RETAILER)”

Therefore the *Backlogged orders (retailer)* variable has been programmed should the warehouse not be able to meet the whole firm order (and neither demand nor backlogged orders). Thus, the final customer's demand forms part of the backlogged orders.

From this point onward, and for simplification purposes, the structure of those variables conditioned by the *IF THEN ELSE* function *(condition, X, Y)* follows this performance; in other words, the result is *X* if the condition is met, otherwise the result is *Y* (this operation is the same for the nested *IF THEN ELSE* functions, that is, those which use several conditioned functions at the same time).

In the constructed model, producing replenishment orders is carried out by considering the *Inventory position (retailer)* variable. This variable informs about the warehouse position at all times and is the result of summing the *Net inventory (retailer)* variable (warehouse position minus any backlogged orders), the on-hold products and deducting the products delivered at this particular time. Therefore:

"INVENTORY ON-HAND (RETAILER)" + "ON ORDER PRODUCTS RETAILER)"—"BACKLOGGED ORDERS (RETAILER)"—PRODUCTS DELIVERED TO FINAL CUSTOMER

The reason for these variables lies in the need to know the warehouse position at all times since the retailer's replenishment orders depend precisely on the warehouse position. So the retailer's replenishment orders sent to the manufacturer are defined by the *Orders to manufacturer* variable:

(LEAD TIME MANUFACTURER*"DEMAND FORECASTING (RETAILER)" + (FILL RATE FACTOR* (SQRT ("VARIANCE DEMAND (RETAILER)"*LEAD TIME MANUFACTURER))))-"INVENTORY POSITION (RETAILER)"

This replenishment order coincides with that explained in the formulation of this problem. As each link depends on the chain members located immediately upstream of it, the real delivery time is not always that foreseen (because of delivery problems, e.g., capacity restrictions at the manufacturer level, which may affect the remaining chain members). The fill rate factor has been considered to equal 2, which statistically corresponds to a fill rate of 97.72%. This percentage may not be reached, or may be exceeded, according to the number of simulations made as the model passes through various stages before it becomes completely stabilized.

The orders sent must be positive, which the *Orders to manufacturer (Positive)* variable controls, which has the following formulation:

MAX (ORDERS TO MANUFACTURER, 0)

In this case the *MAX* order is used so that the system avoids the negative orders that this kind of policy can generate (Sterman 1989). Should there be negative orders, the system response will be null.

Continuing with the analysis of the variables making up the traditional supply chain proposed, the stage is reached that analyzes the finished products arrival process from one level to that located immediately downstream. Thus, the value of the *Orders to manufacturer* variable is sent (information flow toward the manufacturer level) to the *Retailer demand* variable where it is transformed following a process identical to that cited for the retailer case in an order for the manufacturer.

When examining the model, we can see how the result of the *Delivered products to retailer* variable (situated at the manufacturer level, which corresponds to the materials flow) reaches the *Flow of products to retailer* variable, which is associated with the following formulation:

DELIVERED PRODUCTS TO RETAILER/DELAYS 0

This variable is affected by the *Delays* variable, which introduces a delay into the arrival of raw materials for exogenous reasons to the supply chain (in the model, this is assigned the value of 1). This delay can be formulated purely or exponentially. In the previous formulation, it was formulated exponentially.

This materials flow determined by the *Flow of products to retailer* variable feeds the level variable *on-order products (retailer),* which accumulates the units delivered to the retailer from the wholesaler.

This variable's output is the flow variable *Products delivered to retailer,* with the following formulation:

DELAY FIXED (FLOW OF PRODUCTS TO RETAILER, LEAD TIME MANUFACTURER, 0)

As observed, this variable is affected by the *Lead time manufacturer* variable, which introduces a delay into the arrival of raw materials at the retailer warehouse (*Flow of products to retailer).*

The *on-order products (retailer)* variable has been introduced as a level variable since it is necessary to know the amount of on-order products to be delivered to calculate the *Inventory position (retailer)* variable.

The arrival of products at the warehouse takes place exactly after the period defined in the *Lead time manufacturer* variable (provided there are no delays in deliveries caused by possible stockouts); in other words, the delay is pure and not exponential.

The last model level constructed is represented by the manufacturer, which receives the retailer's order commands. As before, if these orders cannot be met with the available inventory, they form part of the backlogged orders. The factory has a daily manufacturing capacity, so it can only manufacture the amount of units on a daily basis which, in terms of the number of hours needed to manufacture each product, the factory is capable of processing.

The information received from the wholesaler level undergoes the same transformations previously cited. In this way, the manufacturer has a sales forecasting which is updated according to the orders received from the immediately previous level.

The service and delivery policies for delayed or backlogged orders follow the same formulation as the previous level and as the retailer level. At this level, delays in the deliveries of orders (lead times) now correspond to the duration of the manufacturing process of those products to be served.

Delivering the orders within the manufacturer's production process follows the patterns detailed below.

The manufacturing capacity is limited by the number of operators available and by the number of hours per period that each operator works. In the model, the *Manufacturer capacity* variable is governed by this formula:

8*NUMBER OF WORKERS/WORKING HOURS PER PRODUCT

Where 8 is the number of hours worked per period (a constant that may be amended) and *Working hours per product* is the variable that determines the number of hours that each product needs to be manufactured. This allows us to obtain the number of units that the manufacturer is able to produce per period.

If the *manufacturing orders* are greater than the factory's daily manufacturing capacity, then the units exceeding this amount are rejected; logically, there is more likelihood of this level incrementing backlogged orders. Therefore, the *Manufacturing constraints* variable follows this formulation:

IF THEN ELSE (PERFORMANCE < 1, "MANUFACTURING ORDERS (POSITIVE)"*PERFORMANCE/DELAYS 1, "MANUFACTURING ORDERS (POSITIVE)"/DELAYS 1)

The conditional function limits the input of orders that must be processed in the factory by means of the *Performance* variable, to which the following formula is assigned:

XIDZ (MANUFACTURING CAPACITY,"MANUFACTURING ORDERS (POSITIVE)", 0)

In other words, the quotient between the *Manufacturing capacity* variable and the *Manufacturing orders (positive)* variable will offer system performance, and if it is greater than or equal to 1, all the orders transfer to the manufacturing, but if below 1, the manufacturing orders are multiplied by system performance (this multiplication always provides the number of units that the system is capable of manufacturing in terms of the *Manufacturing capacity* variable).

The *XIDZ* function prevents the program making an error when the Manufacturing *orders (positive)* variable is 0.

The *Manufacturing constraints* variable is conditioned by the *Delays 1* variable to simulate possible delays due to delays in suppliers' deliveries, machine breakdowns, etc.

The level variable *Manufacturing* accumulates orders by simulating the manufacturing period defined by *Manufacturing lead time* variable. Lastly, the arrival of finished products at the manufacturer warehouse is given by the *Products manufactured* variable.

The constructed model represents the demand management process of a simple traditional supply chain. The reader is recommended, under his or her own criterion, to add new variables for the purpose of adapting the model to any other supply chain type (reduced, e-shopping, EPOS, VMI), to add new levels or to consider multiple products with the help of the Vensim® scripts function, among other extensions. Focusing on diminishing the bullwhip effect, a supply chain model that uses fuzzy estimations in demand instead of exponential smoothing for demand forecasting can be found in Campuzano et al. (2010).

7.4 Conclusions

This chapter brings this short book to an end, which is dedicated to supply chain simulation. Whereas the first chapters in this book (Chaps. 1–4) have centered on highlighting the main theoretical principles to take into account for supply chain simulation for the purpose of improving its performance, the remaining Chaps. 5–7 have done the same, but in a practical manner.

This chapter has considered a simulation model of a traditional supply chain. The problem contemplated, along with its solution, help the reader construct different supply chain models based on the reader's own experience. The systems dynamics models constructed with commercial software (Vensim@ for the examples that this short book provides) enable the reader to simulate by amending the values of the different variables and the various scenarios used, and also help the reader select the one that best adapts to the researcher's, director's and any user's objectives for the company he/she proposes (cutting costs, increasing profits, increasing the fill rate, reducing the number of operators without affecting the manufacturing capacity, etc.).

References

M.S. Amir, Industrial viewpoint. Can systems dynamics be effective in modelling dynamic business systems? Bus. Process. Manage. J. **11**, 612–615 (2005)

F. Campuzano, J. Mula, D. Peidro, Fuzzy estimations and system dynamics for improving supply chains. Fuzzy Set. Syst. **156**, 1530–1542 (2010)

E.A. Silver, D.F. Pyke, R. Peterson, *Inventory Management and Production Planning and Scheduling* (Wiley, New York, 1998)

J.D. Sterman, Modelling managerial behaviour: misperceptions of feedback in a dynamic decision-making experiment. Manage. Sci. **35**, 321–339 (1989)

J.D. Sterman, *Business Dynamics: Systems Thinking and Modeling for a Complex World* (McGraw-Hill Higher Education, New York, 2000)

Index

F. Campuzano and J. Mula, *Supply Chain Simulation*,
DOI: 10.1007/978-0-85729-719-8, © Springer-Verlag London Limited 2011

MIX
Papier aus verantwortungsvollen Quellen
Paper from responsible sources
FSC® C105338

If you have any concerns about our products,
you can contact us on
ProductSafety@springernature.com

In case Publisher is established outside the EU,
the EU authorized representative is:
Springer Nature Customer Service Center GmbH
Europaplatz 3, 69115 Heidelberg, Germany

Printed by Libri Plureos GmbH
in Hamburg, Germany